PREPARING FOR THE

(AB)

AP CALCULUS
EXAMINATION

George Best

J. Richard Lux

Venture Publishing
9 Bartlet Street, Suite 55
Andover, MA 01810

Preface

This workbook is intended for students preparing to take the Advanced Placement Calculus AB Examination. It contains six practice tests that are based on the course description published by the College Board. We have tried to make each of the six tests in this workbook as much like the actual AP Exam as possible. For example, in the appropriate sections, there are questions that require students to make decisions about whether to use the graphing calculator a lot, a little, or not at all. In order to provide a greater supply of this type problem, our exams require the use of a calculator in about half the problems of Section I Part B, and all of Section II Part A.

Each student is expected to have a graphing calculator that has the capability to:
   (1) produce the graph of a function within an arbitrary viewing window,
   (2) find the zeros of a function,
   (3) compute the derivative of a function numerically, and
   (4) compute definite integrals.
In the free-response sections, solutions obtained using one of these four capabilities need only show the setup. Solutions using other calculator capabilities must show the mathematical steps that lead to the answer. In either case, a correct answer alone will not receive full credit.

As in the *AP Course Description for Mathematics*, our examinations are in two sections of equal weight. Section I is all multiple-choice and Section II is all free-response.

1. Section I Part A (28 questions in 55 minutes). Calculators may not be used in this part of the exam.

2. Section I Part B (17 questions in 50 minutes). Calculators are allowed.

3. Section II Part A (3 questions in 45 minutes). Calculators are required

4. Section II Part B (3 questions in 45 minutes). Calculators may not be used and the student may go back to Part A if there is time.

We have tried to create the problems in the spirit of *calculus reform*. Calculus reform implies a change in the mode of instruction as well as increased focus on concepts and less attention to symbolic manipulation; emphasis on modeling and applications; use of technology to explore and deepen understanding of concepts; projects and cooperative learning. We have included questions where functions are defined graphically and numerically, as well as symbolically, in order to give the students more practice in this type of analysis.

We wish to thank the members of the Phillips Academy Mathematics Department for their generous contributions of ideas, problems and advice. Their valuable assistance in testing the problems in the classroom has made us quite confident about the validity of the exams. Robert Clements of Phillips Exeter Academy provided excellent editorial assistance and insightful comments.

In the hope of providing future students with a better workbook, the authors welcome your suggestions, corrections, problems of all sorts, and feedback in general. Please send your comments to:

Venture Publishing
9 Bartlet Street, Suite 55
Andover, Ma 01810
Phone/Fax 508-4896-9486
E-Mail gwbest@tiac.net

George W. Best
J. Richard Lux
Andover, MA

For The Student

There are six examinations in this workbook. Use them as suggested by your teacher, but about two weeks prior to the AP Exam you should try to find a three hour and thirty minute block of time to work through one entire exam. Each part of the exam should be carefully timed. Allow fifty-five minutes for Section I Part A, fifty minutes for Section I Part B, and ninety minutes for Section II. Take a ten minute break between Part A and Part B and also between Part B and Section II. This will give you a good measure of the topics that need more intensive review as well as give you a feel for the energy and enthusiasm needed on a three hour and fifteen minute exam. Repeat the above routine on a second exam four or five days before the AP to check your progress.

The questions on these exams are designed to be as much like the actual AP Exams as possible. However, we have included a greater percentage of medium level and difficult problems and fewer easy ones, in order to help you gain stamina and endurance. If you do a satisfactory job on these exams, then you should be confident of doing well on the actual AP Exam.

The answers to the multiple-choice questions and selected free-response questions are in the back of the workbook. A complete solution manual for all the problems is available from Venture Publishing. No matter how much of an exam you do at one sitting, we strongly urge you to check your answers when you are finished, not as you go along. You will build your confidence if you DO NOT use the "do a problem, check the answer, do a problem" routine.

The following is a list of common student errors:

1. If $f'(c) = 0$, then f has a local maximum or minimum at $x = c$.

2. If $f''(c) = 0$, then the graph of f has an inflection point at $x = c$.

3. If $f'(x) = g'(x)$, then $f(x) = g(x)$.

4. $\dfrac{d}{dx} f(y) = f'(y)$

5. Volume by washers is $\displaystyle\int_a^b (R - r)^2 dx$.

6. Not expressing answers in correct units when units are given.

7. Not providing adequate justification when justification is requested.

8. Wasting time erasing bad solutions. Simply cross out a bad solution after writing the correct solution.

9. Listing calculator results without the supporting mathematics. Recall that a calculator is to be used primarily to:
   a) graph functions,
   b) compute numerical approximations of a derivative and definite integral,
   c) solve equations.

10. Not answering the question that has been asked. For example, if asked to find the maximum value of a function, do not stop after finding the x-value where the maximum value occurs.

Table of Contents

EXAM I

Section I Part A 1
Section I Part B 11
Section II Part A 18
Section II Part A 22

EXAM II

Section I Part A 25
Section I Part B 35
Section II Part A 42
Section II Part A 46

EXAM III

Section I Part A 49
Section I Part B 59
Section II Part A 66
Section II Part A 70

EXAM IV

Section I Part A 73
Section I Part B 83
Section II Part A 90
Section II Part A 94

EXAM V

Section I Part A 97
Section I Part B 107
Section II Part A 114
Section II Part A 118

EXAM VI

Section I Part A 121
Section I Part B 131
Section II Part A 139
Section II Part A 143

ANSWERS 146

INDEX OF TOPICS 149

EXAM I
CALCULUS AB
SECTION I PART A
Time–55 minutes
Number of questions–28

A CALCULATOR MAY NOT BE USED ON THIS PART OF THE EXAMINATION

Directions: Solve each of the following problems, using the available space for scratchwork. After examining the form of the choices, decide which is the best of the choices given and fill in the box. Do not spend too much time on any one problem.

In this test:

(1) Unless otherwise specified, the domain of a function f is assumed to be the set of all real numbers x for which $f(x)$ is a real number.

(2) The inverse of a trigonometric function f may be indicated using the inverse function notation f^{-1} or with the prefix "arc" (e.g., $\sin^{-1} x = \arcsin x$).

---

1. If $f'(x) = \ln(x - 2)$, then the graph of $y = f(x)$ is decreasing if and only if

   (A) $2 < x < 3$ (B) $0 < x$ (C) $0 < x < 1$ (D) $x > 1$ (E) $x > 2$

   Ans

---

2. For $x \neq 0$, the slope of the tangent to $y = x \cos x$ equals zero whenever

   (A) $\tan x = -x$

   (B) $\tan x = \frac{1}{x}$

   (C) $\tan x = x$

   (D) $\sin x = x$

   (E) $\cos x = x$

   Ans

---

3. The function F is defined by

$$F(x) = G[x + G(x)]$$

where the graph of the function G is shown at the right.

The approximate value of $F'(1)$ is

(A) $\frac{7}{3}$

(B) $\frac{2}{3}$

(C) -2

(D) -1

(E) $-\frac{2}{3}$

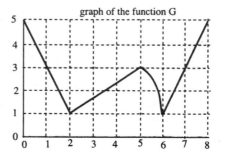

graph of the function G

Ans

4. $\int_{2}^{6} \left(\frac{1}{x} + 2x \right) dx =$

(A) $\ln 4 + 32$

(B) $\ln 3 + 40$

(C) $\ln 3 + 32$

(D) $\ln 4 + 40$

(E) $\ln 12 + 32$

Ans

5. A relative maximum of the function $f(x) = \dfrac{(\ln x)^2}{x}$ occurs at

(A) 0

(B) 1

(C) 2

(D) e

(E) e^2

Ans

6. Use a right-hand Riemann sum with 4 equal subdivisions to approximate the integral
$$\int_{-1}^{3} |2x - 3| \, dx.$$

(A) 13

(B) 10

(C) 8.5

(D) 8

(E) 6

Ans

7. An equation of the line tangent to the graph of $y = x^3 + 3x^2 + 2$ at its point of inflection is

(A) $y = -3x + 1$

(B) $y = -3x - 7$

(C) $y = x + 5$

(D) $y = 3x + 1$

(E) $y = 3x + 7$

Ans

8. $\int \cos(3 - 2x) \, dx =$

(A) $\sin(3 - 2x) + C$

(B) $-\sin(3 - 2x) + C$

(C) $\frac{1}{2}\sin(3 - 2x) + C$

(D) $-\frac{1}{2}\sin(3 - 2x) + C$

(E) $-\frac{1}{5}\sin(3 - 2x) + C$

Ans

9. What is $\displaystyle \lim_{x \to \infty} \frac{\sqrt{9x^2 + 2}}{4x + 3}$?

(A) $\dfrac{3}{2}$ (B) $\dfrac{3}{4}$ (C) $\dfrac{\sqrt{2}}{3}$ (D) 1   (E) The limit does not exist.

Ans ☐

10. Let the first quadrant region enclosed by the graph of $y = \dfrac{1}{x}$ and the lines $x = 1$ and $x = 4$ be the base of a solid. If cross sections perpendicular to the x-axis are semicircles, the volume of the solid is

(A) $\dfrac{3\pi}{64}$

(B) $\dfrac{3\pi}{32}$

(C) $\dfrac{3\pi}{16}$

(D) $\dfrac{3\pi}{8}$

(E) $\dfrac{3\pi}{4}$

Ans ☐

11. Let $f(x) = \ln x + e^{-x}$. Which of the following is TRUE at $x = 1$?

(A) f is increasing

(B) f is decreasing

(C) f is discontinuous

(D) f has a relative minimum

(E) f has a relative maximum

Ans ☐

12.   Let F be the function given by $F(x) = \int_0^x \dfrac{2}{1+t^4}\, dt$. Which of the following statements are true?

   I. $F(0) = 2$
   II. $F(2) < F(6)$
   III. $F''(0) = 0$

   (A) I only   (B) II only   (C) III only   (D) II and III only   (E) I, II, III

   Ans

---

13.   What is the average (mean) value of $2t^3 - 3t^2 + 4$ over the interval $-1 \le t \le 1$?

   (A) 0

   (B) $\dfrac{7}{4}$

   (C) 3

   (D) 4

   (E) 6

   Ans

---

14.   The slope field for a differential equation $\dfrac{dy}{dx} = f(x,y)$ is given in the figure. The slope field corresponds to which of the following differential equations?

   (A) $\dfrac{dy}{dx} = \tan x \cdot \sec x$

   (B) $\dfrac{dy}{dx} = \sin x$

   (C) $\dfrac{dy}{dx} = \cos x$

   (D) $\dfrac{dy}{dx} = -\sin x$

   (E) $\dfrac{dy}{dx} = -\cos x$

   Ans

15. What is $\displaystyle\lim_{x\to 1}\frac{\sqrt{x}-1}{x-1}$?

(A) 0

(B) $\frac{1}{2}$

(C) 1

(D) $\frac{3}{2}$

(E) The limit does not exist.

Ans

☐

16. If $y = \cos^2 x - \sin^2 x$, then $y' =$

(A) -1

(B) 0

(C) $-2(\cos x + \sin x)$

(D) $2(\cos x + \sin x)$

(E) $-4(\cos x)(\sin x)$

Ans

☐

17. The area of the first quadrant region bounded above by the graph of $y = 4x^3 + 6x - \dfrac{1}{x}$ between the values of $x = 1$ and $x = 2$ is

(A) $32 - \ln 2$

(B) $30 - \ln 2$

(C) $24 - \ln 2$

(D) $\dfrac{99}{4}$

(E) 21

Ans

☐

18. $\int \dfrac{x-2}{x-1}\, dx =$

    (A) $-\ln|x-1| + C$

    (B) $x + \ln|x-1| + C$

    (C) $x - \ln|x-1| + C$

    (D) $x + \sqrt{x-1} + C$

    (E) $x - \sqrt{x-1} + C$

Ans

19. Suppose that g is a function with the following two properties: $g(-x) = g(x)$ for all x, and $g'(a)$ exists. Which of the following must necessarily be equal to $g'(-a)$?

    (A) $g'(a)$ (B) $-g'(a)$ (C) $\dfrac{1}{g'(a)}$ (D) $-\dfrac{1}{g'(a)}$ (E) none

Ans

20. An equation for a tangent line to the graph of $y = \text{Arctan}\, \dfrac{x}{3}$ at the origin is:

    (A) $x - 3y = 0$

    (B) $x - y = 0$

    (C) $x = 0$

    (D) $y = 0$

    (E) $3x - y = 0$

Ans

21. If $f(x) = \begin{cases} x^2 + 4 \ \text{ for } \ 0 \le x \le 1 \\ 6 - x \ \text{ elsewhere} \end{cases}$ then $\int\limits_0^3 f(x) \, dx$ is a number between

(A) 0 and 5

(B) 5 and 10

(C) 10 and 15

(D) 15 and 20

(E) 20 and 25

Ans

22. $\dfrac{d}{dx}\left(\ln e^{3x} \right) =$

(A) 1

(B) 3

(C) 3x

(D) $\dfrac{1}{e^{3x}}$

(E) $\dfrac{3}{e^{3x}}$

Ans

23. If $g'(x) = 2g(x)$ and $g(-1) = 1$, then $g(x) =$

(A) e^{2x}

(B) e^{-x}

(C) e^{x+1}

(D) e^{2x+2}

(E) e^{2x-2}

Ans

24. The acceleration at time $t > 0$ of a particle moving along the x-axis is $a(t) = 3t + 2$ ft/sec^2. If at $t = 1$ seconds the velocity is 4 ft/sec and the position is $x = 6$ feet, then at $t = 2$ seconds the position $x(t)$ is

(A) 8 ft   (B) 11 ft   (C) 12 ft   (D) 13 ft   (E) 15 ft

Ans

25. The approximate value of $y = \sqrt{3 + e^x}$ at $x = 0.08$, obtained from the tangent to the graph at $x = 0$, is

(A) 2.01

(B) 2.02

(C) 2.03

(D) 2.04

(E) 2.05

Ans

26. A leaf falls from a tree into a swirling wind. The graph at the right shows its vertical distance (feet) above the ground plotted against time (seconds).

According to the graph, in what time interval is the speed of the leaf the greatest?

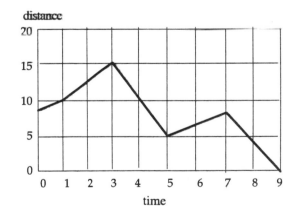

(A) $1 < t < 3$
(B) $3 < t < 5$
(C) $5 < t < 7$
(D) $7 < t < 9$
(E) none of these

Ans

27. Water is flowing into a spherical tank with 6 foot radius at the constant rate of 30π cu ft per hour. When the water is h feet deep, the volume of water in the tank is given by

$$V = \frac{\pi h^2}{3}(18 - h).$$

What is the rate at which the depth of the water in the tank is increasing at the moment when the water is 2 feet deep?

(A) 0.5 ft per hr

(B) 1.0 ft per hr

(C) 1.5 ft per hr

(D) 2.0 ft per hr

(E) 2.5 ft per hr

Ans

28. The graph of the function $f(x) = 2x^{5/3} - 5x^{2/3}$ is increasing on which of the following intervals.

I. $1 < x$ II. $0 < x < 1$ III. $x < 0$

(A) I only   (B) II only   (C) III only   (D) I and II only   (E) I and III only

Ans

EXAM I
CALCULUS AB
SECTION I PART B
Time–50 minutes
Number of questions–17

A GRAPHING CALCULATOR IS REQUIRED FOR SOME QUESTIONS ON THIS PART OF THE EXAMINATION

Directions: Solve each of the following problems, using the available space for scratchwork. After examining the form of the choices, decide which is the best of the choices given and fill in the box. Do not spend too much time on any one problem.

In this test:

(1) The <u>exact</u> numerical value of the correct answer does not always appear among the choices given. When this happens, select from among the choices the number that best approximates the exact numerical value.

(2) Unless otherwise specified, the domain of a function f is assumed to be the set of all real numbers x for which $f(x)$ is a real number.

(3) The inverse of a trigonometric function f may be indicated using the inverse function notation f^{-1} or with the prefix "arc" (e.g., $\sin^{-1} x = \arcsin x$).

---

1. The function f is defined on the interval $[-4, 4]$ and its graph is shown to the right. Which of the following statements are true?

    graph of f

    I. $\lim\limits_{x \to 1} f(x) = -1$

    II. $\lim\limits_{h \to 0} \dfrac{f(2+h) - f(2)}{h} = 2$

    III. $\lim\limits_{x \to -1^+} f(x) = f(-3)$

    (A) I only   (B) II only   (C) I and II only   (D) II and III only   (E) I, II, III

    Ans
    □

---

2. For $f(x) = \sin^2 x$ and $g(x) = 0.5x^2$ on the interval $\left[-\dfrac{\pi}{2}, \dfrac{\pi}{2}\right]$, the instantaneous rate of change of f is greater than the instantaneous rate of change of g for which value of x?

    (A) -0.8 (B) 0 (C) 0.9 (D) 1.2 (E) 1.5

    Ans
    □

---

3. If $f(x) = 2x^2 - x^3$ and $g(x) = x^2 - 2x$, for what values of a and b is

$$\int_a^b f(x)\, dx > \int_a^b g(x)\, dx \ ?$$

   I. $a = -1$ and $b = 0$ II. $a = 0$ and $b = 2$ III. $a = 2$ and $b = 3$

   (A) I only

   (B) II only

   (C) I and II only

   (D) I and III only

   (E) I, II, III

Ans

☐

4. If $y^2 - 3x = 7$, then $\dfrac{d^2y}{dx^2} =$

   (A) $\dfrac{-6}{7y^3}$ (B) $\dfrac{-3}{y^3}$ (C) 3   (D) $\dfrac{3}{2y}$ (E) $\dfrac{-9}{4y^3}$

Ans

☐

5. The graphs of functions f and g are shown at the right.
   If $h(x) = g[f(x)]$, which of the following statements are true about the function h?

   I. $h(0) = 4$.
   II. h is increasing at $x = 2$.
   III. The graph of h has a horizontal tangent at $x = 4$.

   (A) I only   (B) II only   (C) I and II only   (D) II and III only   (E) I, II, III

graph of f

graph of g

Ans

☐

6.   The minimum distance from the origin to the curve $y = e^x$ is

(A) 0.72   (B) 0.74   (C) 0.76   (D) 0.78   (E) 0.80

Ans

7.   The area of the first quadrant region bounded by the y-axis, the line $y = 4 - x$ and the graph of $y = x - \cos x$ is approximately

(A) 4.50   (B) 4.54   (C) 4.56   (D) 4.58   (E) 5.00

Ans

8.   The graph of $y = x^4 - x^2 - e^{2x}$ changes concavity at $x =$

(A) −0.641   (B) −0.531   (C) −0.421   (D) −0.311   (E) −0.201

Ans

9. The rate at which ice is melting in a pond is given by $\dfrac{dV}{dt} = \sqrt{1 + 2^t}$, where V is the volume of ice in cubic feet, and t is the time in minutes. What amount of ice has melted in the first 5 minutes?

(A) 14.49 ft^3 (B) 14.51 ft^3 (C) 14.53 ft^3 (D) 14.55 ft^3 (E) 14.57 ft^3

Ans

10. The region shaded in the figure at the right is rotated about the x-axis. Using the Trapezoid Rule with 5 equal subdivisions, the approximate volume of the resulting solid is

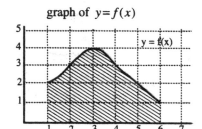

graph of $y = f(x)$

(A) 23

(B) 47

(C) 127

(D) 254

(E) 400

Ans

11. A particle moves along the x-axis so that at time $t \geq 0$, its position is given by $x(t) = (t + 1)(t - 3)^3$. For what values of t is the velocity of the particle increasing?

(A) all t (B) $0 < t < 1$ C) $0 < t < 3$ (D) $1 < t < 3$ E) $t < 1$ or $t > 3$

Ans

12. Let $f(x) = \dfrac{\ln e^{2x}}{x-1}$ for $x > 1$. If g is the inverse of f, then $g'(3) =$

(A) 2   (B) 1   (C) 0   (D) –1   (E) –2

Ans

☐

13. $\displaystyle\int \dfrac{e^{x^2} - 2x}{e^{x^2}}\, dx$

(A) $x - e^{x^2} + C$

(B) $x - e^{-x^2} + C$

(C) $x + e^{-x^2} + C$

(D) $-e^{x^2} + C$

(E) $e^{-x^2} + C$

Ans

☐

14. Suppose f is a function whose derivative is given by $f'(x) = \frac{(x-1)(x-4)^3}{1+x^4}$. Which of the following statements are true?

    I.   The slope of the tangent line to the curve $y = f(x)$ at $x = 2$ is -8.

    II. f is increasing on the interval $(1, 4)$.

    III. f has a local minimum at $x = 4$.

(A) I only   (B) II only   (C) III only   D) II and III only   (E) I, II, III

Ans

15. Let m and b be real numbers and let the function f be defined by
$$f(x) = \begin{cases} 1 + 3bx + 2x^2 & \text{for } x \leq 1 \\ mx + b & \text{for } x > 1. \end{cases}$$
If f is both continuous and differentiable at $x = 1$, then

(A) $m = 1, b = 1$

(B) $m = 1, b = -1$

(C) $m = -1, b = 1$

(D) $m = -1, b = -1$

(E) none of the above

Ans

16. Suppose a car is moving with increasing speed according to the following table.

time (sec)	0	2	4	6	8	10
speed (ft/sec)	30	36	40	48	54	60

The closest approximation of the distance traveled in the first 10 seconds is

(A) 150 ft

(B) 250 ft

(C) 350 ft

(D) 450 ft

(E) 550 ft

Ans

17. Consider the function F defined so that $F(x) + 5 = \int\limits_{2}^{x} \sin\left(\frac{\pi t}{4}\right) dt$.

The value of $F(2) + F'(2)$ is

(A) 0

(B) 1

(C) $\frac{\pi}{4}$

(D) 4

(E) -4

Ans

EXAM I
CALCULUS AB
SECTION II, PART A
Time–45 minutes
Number of problems–3

A graphing calculator is required for some problems or parts of problems.

• Before you begin Part A of Section II, you may wish to look over the problems before starting to work on them. It is not expected that everyone will be able to complete all parts of all problems and you will be able to come back to Part A (without a calculator), if you have time after Part B. All problems are given equal weight, but the parts of a particular solution are not necessarily given equal weight.

• You should write all work for each problem in the space provided. Be sure to write clearly and legibly. If you make an error, you may save time by crossing it out rather than trying to erase it. Erased or crossed out work will not be graded.

• SHOW ALL YOUR WORK. Clearly label any functions, graphs, tables, or other objects you use. You will be graded on the correctness and completeness of your methods as well as your final answers. Answers without supporting work may not receive credit.

• Justifications require that you give mathematical (noncalculator) reasons.

• You are permitted to use your calculator in Part A to solve an equation, find the derivative of a function at a point, or calculate the value of a definite integral. However, you must clearly indicate in your exam booklet the setup of your problem, namely the equation, function, or integral you are using. If you use other built-in features or programs, you must show the mathematical steps necessary to produce your results.

• Your work must be expressed in mathematical notation rather than calculator syntax. For example,

$$\int_1^5 x^2 \, dx \quad \text{may not be written as} \quad \text{fnInt}(X^2, X, 1, 5).$$

• Unless otherwise specified, answers (numeric or algebraic) need not be simplified.

• If you use decimal approximations in your calculations, you will be graded on accuracy. Unless otherwise specified, your final answers should be accurate to three places after the decimal point.

• Unless otherwise specified, the domain of a function f is assumed to be the set of all real numbers x for which $f(x)$ is a real number.

THE EXAM BEGINS ON THE NEXT PAGE

PLEASE TURN OVER

1.   Two functions, *f* and *g* , are defined on the closed interval $-4 \le x \le 4$. A graph of the
      function *f* is given in the following figure.

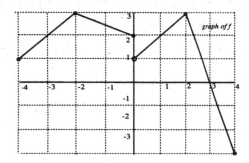

      The table below contains some values of the continuous function *g*.

x	-4	-3	-2	-1	0	1	2	3	4
g(x)	10	9	5	-1	0	2	6	0	-3

(a) Find $f'(3)$.

(b) Approximate $g'(0)$. Show your work.

(c) If the function *h* is defined by $h(x) = g[f(x)]$, evaluate: i) $h(2)$ and ii) $h'(3)$

(d) Approximate $\displaystyle\int_0^4 f(x)\, dx$

2.   Let f be the function given by $f(x) = x^3 + 3x^2 - x + 2$.

   (a) The tangent to the graph of f at the point $P = (-2, 8)$ intersects the graph of f again at the point Q. Find the coordinates of the point Q.

   (b) Find the coordinates of point R, the inflection point on the graph of f.

   (c) Show that the segment $\overline{QR}$ divides the region between the graph of f and its tangent at P into two regions whose areas are in the ratio of $\frac{16}{11}$.

3.   Consider the graphs of $y = 3x + c$ and $y^2 = 6x$, where c is a real constant.

(a)   Determine all values of c for which the graphs intersect in two distinct points.

(b)   Suppose $c = -\dfrac{3}{2}$. Find the area of the region enclosed by the two curves.

(c)   Suppose $c = 0$. Find the volume of the solid formed when the region bounded by $y = 3x$ and $y^2 = 6x$ is revolved about the x-axis.

A CALCULATOR MAY **NOT** BE USED ON THIS PART OF THE EXAMINATION.
DURING THE TIMED PORTION FOR PART B, YOU MAY GO BACK AND CONTINUE TO WORK
ON THE PROBLEMS IN PART A WITHOUT THE USE OF A CALCULATOR.

4.   Consider the differential equation $\frac{dy}{dx} = x - y$.

   (a) On the axes provided, sketch a slope field for the given differential equation at the
       fourteen points indicated.

   (b) Sketch the solution curve that contains the point $(-1, 1)$.

   (c) Find an equation for the straight line solution through the point $(1, 0)$.

   (d) Show that if C is a constant, then $y = x - 1 + Ce^{-x}$ is a solution of the differential
       equation.

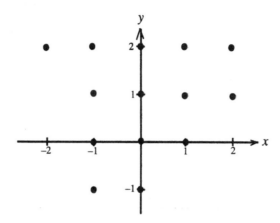

5. Let *f* be a function defined on the closed interval $-4 \leq x \leq 4$.
    The graph of f', the **derivative** of f, is shown in the
    figure. The graph of f' has horizontal tangents at
    $x = -3, -1$ and 3.

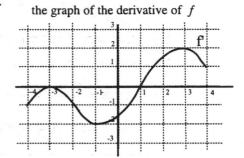

the graph of the derivative of *f*

(a) Find an equation of the line tangent to the graph of f
     at the point (3, 1).

(b) Find all values of x on the open $(-4, 4)$ at which f has
     a local minimum? Justify your answer.

(c) Estimate $f''(2)$.

(d) Find the x-coordinate of each point of inflection of the graph of f on the open interval
     $(-4, 4)$. Justify your answer.

(e) At what values of x does f achieve its maximum on the closed interval [0, 4]?

6.   Let A be the area of the region in the first quadrant under the graph of $y = \cos x$ and above the line $y = k$ for $0 \le k \le 1$.

   (a) Determine A in terms of k.

   (b) Determine the value of A when $k = \dfrac{1}{2}$.

   (c) If the line $y = k$ is moving upward at the rate of $\dfrac{1}{\pi}$ units per minute, at what rate is the area, A, changing when $k = \dfrac{1}{2}$?

EXAM II
CALCULUS AB
SECTION I PART A
Time–55 minutes
Number of questions–28

A CALCULATOR MAY NOT BE USED ON THIS PART OF THE EXAMINATION

Directions: Solve each of the following problems, using the available space for scratchwork. After examining the form of the choices, decide which is the best of the choices given and fill in the box. Do not spend too much time on any one problem.

In this test:

(1) Unless otherwise specified, the domain of a function f is assumed to be the set of all real numbers x for which $f(x)$ is a real number.

(2) The inverse of a trigonometric function f may be indicated using the inverse function notation f^{-1} or with the prefix "arc" (e.g., $\sin^{-1} x = \arcsin x$).

---

1.   Let $f(x) = 4x^3 - 3x - 1$. An equation of the line tangent to $y = f(x)$ at $x = 2$ is

(A) $y = 25x - 5$

(B) $y = 45x + 65$

(C) $y = 45x - 65$

(D) $y = 65 - 45x$

(E) $y = 65x - 45$

Ans

---

2. $\displaystyle\int_0^1 \sin \pi x \; dx =$

(A) $\dfrac{2}{\pi}$ (B) $\dfrac{1}{\pi}$ (C) 0   (D) $-\dfrac{2}{\pi}$ (E) $-\dfrac{1}{\pi}$

Ans

---

3. $\lim\limits_{h \to 0} \left(\dfrac{\cos(x + h) - \cos x}{h} \right) =$

(A) $\sin x$

(B) $- \sin x$

(C) $\cos x$

(D) $- \cos x$

(E) does not exist

Ans

4. On which of the following intervals, is the graph of the curve $y = x^5 - 5x^4 + 10x + 15$ concave up ?

I. $x < 0$ II. $0 < x < 3$ III. $x > 3$

(A) I only

(B) II only

(C) III only

(D) I and II only

(E) II and III only

Ans

5. The region bounded by the x-axis and the part of the graph of $y = \sin x$ between $x = 0$ and $x = \pi$ is separated into two regions by the line $x = k$. If the area of the region for $0 \le x \le k$ is one-third the area of the region for $k \le x \le \pi$, then $k =$

(A) $\arcsin \dfrac{1}{3}$

(B) $\arcsin \dfrac{1}{4}$

(C) $\dfrac{\pi}{6}$

(D) $\dfrac{\pi}{3}$

(E) $\dfrac{\pi}{4}$

Ans

6. A particle starts at time $t = 0$ and moves along a number line so that its position, at time $t \geq 0$, is given by $x(t) = (t-2)^3(t-6)$. The particle is moving to the right for

   (A) $0 < t < 5$

   (B) $2 < t < 6$

   (C) $t > 5$

   (D) $t \geq 0$

   (E) never

   Ans
   □

7. An antiderivative of $(x^2 - 1)^2$ is

   (A) $\frac{1}{3}(x^2 - 1)^3 + C$

   (B) $\frac{1}{5}x^5 - x + C$

   (C) $4x(x^2 - 1) + C$

   (D) $\frac{1}{6x}(x^2 - 1)^3 + C$

   (E) $\frac{1}{5}x^5 - \frac{2}{3}x^3 + x + C$

   Ans
   □

8. $\displaystyle\int_{\pi/4}^{\pi/3} \frac{\sec^2 x}{\tan x}\, dx =$

   (A) $\ln\sqrt{3}$ (B) $-\ln\sqrt{3}$ (C) $\ln\sqrt{2}$ (D) $\sqrt{3}-1$ (E) $\ln\frac{\pi}{3} - \ln\frac{\pi}{4}$

   Ans
   □

9.   What is $\displaystyle\lim_{x \to \infty} \frac{x^2 - 6}{2 + x - 3x^2}$?

(A) -3 (B) $-\dfrac{1}{3}$ (C) $\dfrac{1}{3}$ (D) 2 (E) The limit does not exist.

Ans

10. $\displaystyle\int_0^2 \sqrt{x^2 - 4x + 4}\ dx$ is:

(A) 1

(B) -1

(C) -2

(D) 2

(E) None of the above

Ans

11.   If $g(x) = \dfrac{x - 2}{x + 2}$, then $g'(2) =$

(A) 1

(B) -1

(C) $\dfrac{1}{4}$

(D) $-\dfrac{1}{4}$

(E) 0

Ans

12. Which of the following is the solution to the differential equation $\dfrac{dy}{dx} = y^2$, where $y(-1) = 1$?

(A) $y = \frac{1}{x}$ for $x \neq 0$

(B) $y = -\frac{1}{x}$ for $x < 0$

(C) $y = -\frac{1}{x}$ for $x > 0$

(D) $y = \frac{1}{x}$ for $x > 0$

(E) $y = \frac{1}{x}$ for $x < 0$

Ans

13. The fourth derivative of $f(x) = (2x - 3)^4$ is

(A) $24(2^4)$

(B) $24(2^3)$

(C) $24(2x - 3)$

(D) $24(2^5)$

(E) 0

Ans

14. If $\displaystyle\int_2^4 f(x)\,dx = 6$, then $\displaystyle\int_2^4 (f(x) + 3)\,dx =$

(A) 3

(B) 6

(C) 9

(D) 12

(E) 15

Ans

15. The slope of the tangent line to the curve $2xy + \sin y = 2\pi$ at the point where $y = \pi$ is

(A) -2π

(B) $-\pi$

(C) 0

(D) π

(E) 2π

Ans

16. If $f(x) = e^{2x}$ and $g(x) = \ln x$, then the derivative of $y = f(g(x))$ at $x = e$ is

(A) e^2

(B) $2e^2$

(C) $2e$

(D) 2

(E) undefined

Ans

17. The area of the region bounded by the lines $x = 1$ and $y = 0$ and the curve $y = xe^{x^2}$ is

(A) $1 - e$

(B) $e - 1$

(C) $\dfrac{e - 1}{2}$

(D) $\dfrac{1 - e}{2}$

(E) $\dfrac{e}{2}$

Ans

18.   If $h(x) = (x^2 - 4)^{3/4} + 1$, then the value of $h'(2)$ is

     (A) 3

     (B) 2

     (C) 1

     (D) 0

     (E) does not exist

Ans

19.   The derivative of $\sqrt{x} - \dfrac{1}{x\sqrt[3]{x}}$ is

     (A) $\dfrac{1}{2} x^{-1/2} - x^{-4/3}$

     (B) $\dfrac{1}{2} x^{-1/2} + \dfrac{4}{3} x^{-7/3}$

     (C) $\dfrac{1}{2} x^{-1/2} - \dfrac{4}{3} x^{-1/3}$

     (D) $-\dfrac{1}{2} x^{-1/2} + \dfrac{4}{3} x^{-7/3}$

     (E) $-\dfrac{1}{2} x^{-1/2} - \dfrac{4}{3} x^{-1/3}$

Ans

20.   The function f is continuous at $x = 1$.

$$\text{If} \quad f(x) = \begin{cases} \dfrac{\sqrt{x+3} - \sqrt{3x+1}}{x-1} & \text{for } x \neq 1 \\[2ex] k & \text{for } x = 1 \end{cases}$$

                                             then $k =$

     (A) 0   (B) 1   (C) $\dfrac{1}{2}$ (D) $-\dfrac{1}{2}$ (E) none of the above

Ans

21.   An equation of the normal to the graph of $f(x) = \dfrac{x}{2x-3}$ at $(1, f(1))$ is

(A) $3x + y = 4$

(B) $3x + y = 2$

(C) $x - 3y = -2$

(D) $x - 3y = 4$

(E) $x + 3y = 2$

Ans

22.   Let $f(x) = x \ln x$. The minimum value attained by f is

(A) $-\dfrac{1}{e}$

(B) 0

(C) $\dfrac{1}{e}$

(D) -1

(E) There is no minimum.

Ans

23. The slope field for a differential equation $\frac{dy}{dx} = f(x, y)$ is given in the figure. The slope field corresponds to which of the following differential equations?

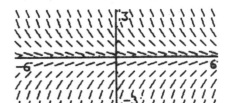

   (A) $\frac{dy}{dx} = x + y$

   (B) $\frac{dy}{dx} = y^2$

   (C) $\frac{dy}{dx} = -y$

   (D) $\frac{dy}{dx} = e^{-x}$

   (E) $\frac{dy}{dx} = 1 - \ln x$

Ans

24. The average value of $\sec^2 x$ over the interval $0 \le x \le \frac{\pi}{4}$ is

   (A) $\frac{\pi}{4}$ (B) $\frac{4}{\pi}$ (C) $\frac{\pi}{8}$ (D) 1   (E) none of the above

Ans

25. Suppose that g is a function that is defined for all real numbers. Which of the following conditions assures that g has an inverse function?

   (A) $g'(x) < 1$, for all x

   (B) $g'(x) > 1$, for all x

   (C) $g''(x) > 0$, for all x

   (D) $g''(x) < 0$, for all x

   (E) g is continuous.

Ans

26. The function f is continuous and differentiable on the closed interval $[1, 5]$. The table below gives selected values of f on this interval. Which of the following statements must be TRUE?

x	1	2	3	4	5
$f(x)$	3	4	5	3	-2

(A) $f'(x) > 0$ for $1 < x < 3$

(B) $f''(x) < 0$ for $3 < x < 5$

(C) The maximum value of f on $[1, 5]$ must be 5.

(D) The minimum value of f on $[1, 5]$ must be -2.

(E) There exists a number c, $1 < c < 5$ for which $f(c) = 0$.

Ans ☐

27. If the function G is defined for all real numbers by $G(x) = \int_0^{2x} \cos(t^2)\, dt$, then $G'(\sqrt{\pi}) =$

(A) 2   (B) 1   (C) 0   (D) –1   (E) –2

Ans ☐

28. At time t a particle moving along the x-axis is at position x. The relationship between x and t is given by: $tx = x^2 + 8$. At $x = 2$ the velocity of the particle is

(A) 1

(B) 2

(C) 6

(D) –2

(E) –1

Ans ☐

EXAM II
CALCULUS AB
SECTION I PART B
Time–50 minutes
Number of questions–17

A GRAPHING CALCULATOR IS REQUIRED FOR SOME QUESTIONS ON THIS PART OF THE EXAMINATION

Directions: Solve each of the following problems, using the available space for scratchwork. After examining the form of the choices, decide which is the best of the choices given and fill in the box. Do not spend too much time on any one problem.

In this test:

(1)   The exact numerical value of the correct answer does not always appear among the choices given. When this happens, select from among the choices the number that best approximates the exact numerical value.

(2)   Unless otherwise specified, the domain of a function f is assumed to be the set of all real numbers x for which $f(x)$ is a real number.

(3)   The inverse of a trigonometric function f may be indicated using the inverse function notation f^{-1} or with the prefix "arc" (e.g., $\sin^{-1} x = \arcsin x$).

---

1.   Which of the following functions have a derivative at $x = 0$?

   I. $y = \left| x^3 - 3x^2 \right|$

   II. $y = \sqrt{x^2 + .01} - \left| x - 1 \right|$

   III. $y = \dfrac{e^x}{\cos x}$

   (A) None   (B) II only   (C) III only   (D) II and III only   (E) I, II, III

Ans ☐

---

2.   Water is pumped into an empty tank at a rate of $r(t) = 20e^{0.02t}$ gallons per minute. Approximately how many gallons of water have been pumped into the tank in the first five minutes?

   (A) 20 gal

   (B) 22 gal

   (C) 85 gal

   (D) 105 gal

   (E) 150 gal

Ans ☐

---

3.   Consider the function $f(x) = \dfrac{6x}{a + x^3}$ for which $f'(0) = 3$. The value of a is

(A) 5

(B) 4

(C) 3

(D) 2

(E) 1

Ans

☐

---

4.   Which of the following is true about the function f if $f(x) = \dfrac{(x-1)^2}{2x^2 - 5x + 3}$?

I. f is continuous at $x = 1$.

II.   The graph of f has a vertical asymptote at $x = 1$.

III.   The graph of f has a horizontal asymptote at $y = \dfrac{1}{2}$.

(A) I only   (B) II only   (C) III only   (D) II and III only   (E) I, II, III

Ans

☐

---

5.   If $y = u + 2e^u$ and $u = 1 + \ln x$, find $\dfrac{dy}{dx}$ when $x = \dfrac{1}{e}$

(A) e (B) $2e$ (C) $3e$ (D) $\dfrac{2}{e}$ (E) $\dfrac{3}{e}$

Ans

☐

---

6.   Sand is being dumped on a pile in such a way that it
      always forms a cone whose base radius is always 3
      times its height. The function V whose graph is
      sketched in the figure gives the volume of the
      conical sand pile, $V(t)$, measured in cubic feet, after

      t minutes. $\left(V(t) = \dfrac{1}{3}\pi r^2 h \right)$ At what approximate

      rate is the radius of the base changing after 6
      minutes?

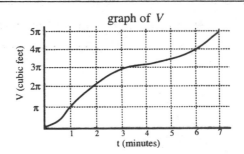

graph of V

      (A) 0.22 ft/min   (B) 0.28 ft/min   (C) 0.34 ft/min   (D) 0.40 ft/min   (E) 0.46 ft/min

Ans

7.   A graph of the function f is shown at the right.
      Which of the following statements are true?

      I. $f(1) > f'(3)$

      II. $\displaystyle\int_{1}^{2} f(x)\,dx > f'(3.5)$

      III. $\displaystyle\lim_{h \to 0} \dfrac{f(2+h)-f(2)}{h} > \dfrac{f(2.5)-f(2)}{2.5-2}$

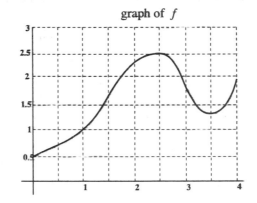

graph of f

      (A) I only   (B) II only   (C) I and II only   (D) II and III only   (E) I, II, III

Ans

8.   Given: $5x^3 + 40 = \displaystyle\int_{a}^{x} f(t)\,dt$. The value of a is

      (A) –2
      (B) 2
      (C) 1
      (D) –1
      (E) 0

Ans

9.   Let R be the region in the first quadrant enclosed by the lines $x = \ln 3$ and $y = 1$ and the graph of $y = e^{x/2}$. The volume of the solid generated when R is revolved about the line $y = -1$ is

    (A) 5.128

    (B) 7.717

    (C) 12.845

    (D) 15.482

    (E) 17.973

Ans

10. If $\dfrac{dy}{dx} = \dfrac{x\sin(x^2)}{y}$, then y could be

    (A) $\sqrt{2 - \cos(x^2)}$

    (B) $\sqrt{2} - \cos(x^2)$

    (C) $2 - \cos(x^2)$

    (D) $\cos(x^2)$

    (E) $\sqrt{2 - \cos x}$

Ans

11. If $y = \sin u$, $u = v - \dfrac{1}{v}$, and $v = \ln x$, then value of $\dfrac{dy}{dx}$ at $x = e$ is

    (A) 0

    (B) 1

    (C) $\dfrac{1}{e}$

    (D) $\dfrac{2}{e}$

    (E) $\cos e$

Ans

12.   The area of the region bounded by the graphs of $y = 2 - |x - 3|$ and $y = x^2 - 2x$ is

 (A) 1.86   (B) 1.88   (C) 1.90   (D) 1.92   (E) 1.94

Ans

☐

---

13.   The figure shows the graph of f', the   graph of the derivative of f
       derivative of a function f. The domain of f
       is the interval $-4 \le x \le 4$. Which of the
       following are true about the graph of f?

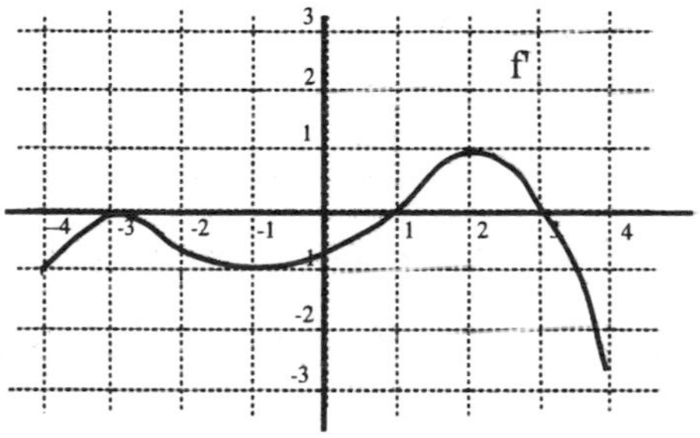

   I.   At the points where $x = -3$ and $x = 2$
        there are horizontal tangents.

  II.   At the point where $x = 1$ there is a relative
        minimum point.

 III.   At the point where $x = -3$ there is an inflection point.

   (A) None   (B) II only   (C) III only   (D) II and III only   (E) I, II, III

Ans

☐

14. A differentiable function f has the property that $f(3) = 5$ and $f'(3) = 4$. What is an estimate for $f(2.8)$ using the linear appoximation for f at $x = 3$?

    (A) 6.6

    (B) 5.8

    (C) 5.0

    (D) 4.2

    (E) 3.4

Ans
☐

15. The number of bacteria in a culture is given by $N(t) = 200\ln(t^2 + 36)$, where t is measured in days. On what day is the change in growth a maximum?

    (A) 4   (B) 6   (C) 8   (D) 10   (E) 12

Ans
☐

16. The acceleration of a particle at time t moving along the x-axis is given by: $a = 4e^{2t}$.
    At the instant when $t = 0$, the particle is at the point $x = 2$ moving with velocity $v = -2$.
    The position of the particle at $t = \frac{1}{2}$ is

    (A) $e - 3$ (B) $e - 2$ (C) $e - 1$ (D) e (E) $e + 1$

    Ans ⬜

17. The graph of f', the derivative of a
    function f, is shown at the right. The graph
    of f', has a horizontal tangent at $x = 0$.

    Which of the following statements are true
    about the function f?

    I. f is increasing on the interval $(-2, -1)$.
    II. f has an inflection point at $x = 0$.
    III. f is concave up on the interval $(-1, 0)$.

    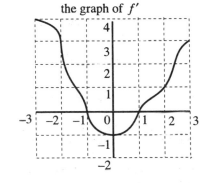
    the graph of f'

    (A) I only   (B) II only   (C) III only   (D) I and II only   (E) II and III only

    Ans ⬜

EXAM II
CALCULUS AB
SECTION II, PART A
Time–45 minutes
Number of problems–3

A graphing calculator is required for some problems or parts of problems.

- Before you begin Part A of Section II, you may wish to look over the problems before starting to work on them. It is not expected that everyone will be able to complete all parts of all problems and you will be able to come back to Part A (without a calculator), if you have time after Part B. All problems are given equal weight, but the parts of a particular solution are not necessarily given equal weight.

- You should write all work for each problem in the space provided. Be sure to write clearly and legibly. If you make an error, you may save time by crossing it out rather than trying to erase it. Erased or crossed out work will not be graded.

- SHOW ALL YOUR WORK. Clearly label any functions, graphs, tables, or other objects you use. You will be graded on the correctness and completeness of your methods as well as your final answers. Answers without supporting work may not receive credit.

- Justifications require that you give mathematical (noncalculator) reasons.

- You are permitted to use your calculator in Part A to solve an equation, find the derivative of a function at a point, or calculate the value of a definite integral. However, you must clearly indicate in your exam booklet the setup of your problem, namely the equation, function, or integral you are using. If you use other built-in features or programs, you must show the mathematical steps necessary to produce your results.

- Your work must be expressed in mathematical notation rather than calculator syntax. For example, $\int_1^5 x^2 \, dx$ may not be written as fnInt(X^2, X, 1, 5).

- Unless otherwise specified, answers (numeric or algebraic) need not be simplified.

- If you use decimal approximations in your calculations, you will be graded on accuracy. Unless otherwise specified, your final answers should be accurate to three places after the decimal point.

- Unless otherwise specified, the domain of a function f is assumed to be the set of all real numbers x for which $f(x)$ is a real number.

THE EXAM BEGINS ON THE NEXT PAGE

PLEASE TURN OVER

1.   Let R be the shaded region in the first
     quadrant enclosed by the *y*-axis and the graphs
     of $y = 4 - x^2$ and $y = 1 + 2 \sin x$ as shown
     in the figure at the right.

     (a) Find the area of R.

     (b) Find the volume of the solid generated
          when R is revolved about the *x*-axis.

     (c) Find the volume of the solid whose base
          is R and whose cross sections
          perpendicular to the *x*-axis are squares.

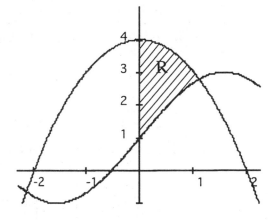

2. A square is inscribed in a circle as shown in the figure at the right. As the square expands, the circle expands to maintain the four points of intersection. The perimeter of the square is increasing at the rate of 8 inches per second.

    (For the circle: $A = \pi r^2$ and $C = 2\pi r$.)

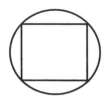

    (a) Find the rate at which the circumference of the circle is increasing.

    (b) At the instant when the area of the square is 16 square inches, find the rate at which the area enclosed between the square and the circle is increasing.

3.   Suppose the function F is defined by $F(x) = \int\limits_1^{\sqrt{x}} \dfrac{2t-1}{t+2}\, dt$ for all real numbers $x \geq 0$.

(a) Evaluate $F(1)$.

(b) Evaluate $F'(1)$

(c) Find an equation for the tangent line to the graph of F at the point where $x = 1$.

(d) On what intervals is the function F increasing? Justify your answer.

A CALCULATOR MAY **NOT** BE USED ON THIS PART OF THE EXAMINATION.
DURING THE TIMED PORTION FOR PART B, YOU MAY GO BACK AND CONTINUE TO WORK
ON THE PROBLEMS IN PART A WITHOUT THE USE OF A CALCULATOR.

4.   Suppose that a population of bacteria grows according to the logistic equation
$\frac{dP}{dt} = 2P(3 - P)$, where P is the population measured in thousands and t is time measured in days. A slope field for this equation is given below.

(a) Sketch the solution curve that passes through the point $(0, \frac{1}{10})$ and sketch the solution curve that passes through the point $(0, 4)$. Which solution has an inflection point?

(b) Suppose the bacteria population began at day 0 with 1000 members, that is $P(0) = 1$. Find an equation of the line tangent to the solution curve $y = P(t)$ at the point $(0, 1)$.

(c) Show that the function $P(t) = \frac{3}{1 + 2e^{-6t}}$ is a solution of the differential equation.

(d) Show that the maximum growth rate of the bacteria occurs at $P = \frac{3}{2}$.

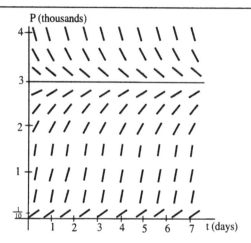

5.   A function F is defined on the closed
     interval $-3 \le x \le 4$. The graph of F', the
     derivative of F, is shown at the right.
     The graph of F' has horizontal tangents at
     $x = -2$ and 2.

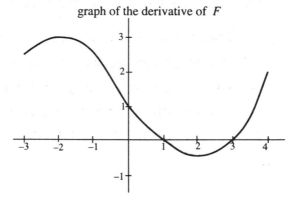

graph of the derivative of F

   (a) On what intervals, if any, is F
        increasing. Justify your answer.

   (b) At what value of x does F attain its
        absolute maximum value on the closed
        interval $[-3, 4]$. Show the analysis that
        leads to your answer.

   (c) Find the interval(s) for which the graph
        of F is concave down.

6. Let $f(x) = ax + \dfrac{b}{x}$ where a and b are positive constants.

    (a) Find in terms of a and b, the intervals on which f is increasing.

    (b) Find the coordinates of all local maximum and minimum points.

    (c) On what interval(s) is the graph concave up?

    (d) Find any inflection points. Explain briefly.

EXAM III
CALCULUS AB
SECTION I PART A
Time–55 minutes
Number of questions–28

A CALCULATOR MAY NOT BE USED ON THIS PART OF THE EXAMINATION

Directions: Solve each of the following problems, using the available space for scratchwork. After examining the form of the choices, decide which is the best of the choices given and fill in the box. Do not spend too much time on any one problem.

In this test:

(1) Unless otherwise specified, the domain of a function f is assumed to be the set of all real numbers x for which $f(x)$ is a real number.

(2) The inverse of a trigonometric function f may be indicated using the inverse function notation f^{-1} or with the prefix "arc" (e.g., $\sin^{-1}x = \arcsin x$).

---

1. Which of the following is a function with a vertical tangent at $x = 0$?

   (A) $f(x) = x^3$ (B) $f(x) = \sqrt[3]{x}$ (C) $f(x) = \dfrac{1}{x}$ (D) $f(x) = \sin x$ (E) $f(x) = \tan x$

   Ans ☐

---

2. $\displaystyle\int_0^5 \frac{dx}{\sqrt{1 + 3x}} =$

   (A) 4   (B) $\dfrac{8}{3}$ (C) 2   (D) $\dfrac{16}{9}$ (E) 1

   Ans ☐

---

3.   Which function is NOT continuous everywhere?

   (A) $y = |x|$

   (B) $y = x^{2/3}$ look A domain

   (C) $y = \sqrt{x^2 + 1}$

   (D) $y = \dfrac{x}{x^2 + 1}$

   ✗ (E) $y = \dfrac{4x}{(x + 1)^2}$

Ans

☐

4.   The area of the first quadrant region bounded by the curve $y = e^{-x}$, the x-axis, the y-axis and the line $x = 2$ is equal to

   (A) 1

   (B) 2

   (C) $\ln e^x$

   (D) $\dfrac{1}{e^2} - 1$

   (E) $1 - \dfrac{1}{e^2}$

Ans

☐

5.   If $g(x) = x + \cos x$, then $\displaystyle\lim_{h \to 0} \dfrac{g(x + h) - g(x)}{h} =$

   Definition of Deri.

   (A) $\sin x + \cos x$

   (B) $\sin x - \cos x$

   (C) $1 - \sin x$

   (D) $1 - \cos x$

   (E) $x^2 - \sin x$

Ans

☐

6. $\displaystyle\int_{0}^{4} \frac{2x}{x^2+9}\,dx =$

(A) 25

(B) 16

(C) $\ln\dfrac{25}{9}$

(D) $\ln 4$

(E) $\ln\dfrac{5}{3}$

Ans

7. A function g is defined for all real numbers and has the following property:
$g(a+b) - g(a) = 4ab + 2b^2$. Find $g'(x)$.

(A) 4

(B) -4

(C) $2x^2$

(D) $4x$

(E) does not exist

Ans

8. Given the function defined by $f(x) = x^5 - 5x^4 + 3$, find all values of x for which the graph of f is concave up.

(A) $x > 0$

(B) $x > 3$

(C) $0 < x < 3$

(D) $x < 0$ or $x > 3$

(E) $x < 0$ or $x > 5$

Ans

9.   If $f(x) = 2 + |x|$, find the average value of the function f on the interval $-1 \le x \le 3$.

   (A) $\dfrac{7}{4}$ (B) $\dfrac{9}{4}$ (C) $\dfrac{11}{4}$ (D) $\dfrac{13}{4}$ (E) $\dfrac{15}{4}$

                                                                                        Ans

10.   A particle starts at $(5, 0)$ when $t = 0$ and moves along the x-axis in such a way that at

      time $t > 0$ its velocity is given by $v(t) = \dfrac{1}{1 + t}$. Determine the position of the particle

      at $t = 3$.

   (A) $\dfrac{97}{16}$

   (B) $\dfrac{95}{16}$

   (C) $\dfrac{79}{16}$

   (D) $1 + \ln 4$

   (E) $5 + \ln 4$

                                                                                        Ans

11.   If $g(x) = \sqrt[3]{x - 1}$ and f is the inverse function of g, then $f'(x) =$

   (A) $3x^2$

   (B) $3(x - 1)^2$

   (C) $-\dfrac{1}{3}(x - 1)^{-4/3}$

   (D) $\dfrac{1}{3}(x - 1)^{2/3}$

   (E) does not exist

                                                                                        Ans

12. Suppose $F(x) = \int\limits_0^{\cos x} \sqrt{1 + t^3}\; dt$ for all real x, then $F'\left(\dfrac{\pi}{2}\right) =$

(A) -1

(B) 0

(C) $\dfrac{1}{2}$

(D) 1

(E) $\dfrac{\sqrt{3}}{2}$

Ans ☐

13. If the line $3x - y + 2 = 0$ is tangent in the first quadrant to the curve $y = x^3 + k$, then $k =$

(A) 5

(B) -5

(C) 4

(D) 1

(E) -1

Ans ☐

14. The slope field for a differential equation $\dfrac{dy}{dx} = f(x, y)$ is given in the figure. Which of the following statements are true?

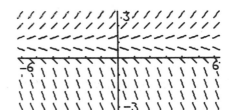

I. A solution curve that contains the point $(0, 2)$ also contains the point $(-2, 0)$.

II. As y approaches 1, the rate of change of y approaches zero.

III. All solution curves for the differential equation have the same slope for a given value of y

(A) I only   (B) II only   (C) I and II only   (D) II and III only   (E) I, II, III

Ans ☐

15. $\dfrac{d}{dx}[\text{Arctan } 3x] =$

(A) $\dfrac{1}{1+9x^2}$

(B) $\dfrac{3}{1+9x^2}$

(C) $\dfrac{3}{\sqrt{4x^2-1}}$

(D) $\dfrac{3}{1+3x}$

(E) none of the above

Ans

☐

16. $\displaystyle\lim_{x\to 1} \dfrac{x^2+2x-3}{x^2-1} =$

(A) -2

(B) -1

(C) 10

(D) 1

(E) 2

Ans

☐

17. If f and g are continuous functions such that $g'(x) = f(x)$ for all x, then $\displaystyle\int_2^3 f(x)\, dx =$

(A) $g'(2) - g'(3)$
(B) $g'(3) - g'(2)$
(C) $g(3) - g(2)$
(D) $f(3) - f(2)$
(E) $f'(3) - f'(2)$

Ans

☐

18. Let $y = 2e^{\cos x}$. Both x and y vary with time in such a way that y increases at the constant rate of 5 units per second. The rate at which x is changing when $x = \frac{\pi}{2}$ is

(A) 10 units/sec

(B) −10 units/sec

(C) −2.5 units/sec

(D) 2.5 units/sec

(E) −0.4 units/sec

Ans

19. $\displaystyle\int_1^2 \frac{dx}{x^3} =$

(A) $\frac{3}{8}$

(B) $-\frac{3}{8}$

(C) $\frac{15}{64}$

(D) $\frac{3}{4}$

(E) $\frac{15}{16}$

Ans

20. The maximum distance, measured horizontally, between the graphs of $f(x) = x$ and $g(x) = x^2$ for $0 \le x \le 1,$ is

(A) 1   (B) $\frac{3}{4}$ (C) $\frac{1}{2}$ (D) $\frac{1}{4}$ (E) $\frac{1}{8}$

Ans

21.   Let f be the function defined by $f(x) = \begin{cases} x+1 & \text{for } x < 0 \\ 1 + \sin \pi x & \text{for } x \geq 0. \end{cases}$ Then $\int_{-1}^{1} f(x)\, dx =$

(A) $\dfrac{3}{2}$

(B) $\dfrac{3}{2} - \dfrac{2}{\pi}$

(C) $\dfrac{1}{2} - \dfrac{2}{\pi}$

(D) $\dfrac{3}{2} + \dfrac{2}{\pi}$

(E) $\dfrac{1}{2} + \dfrac{2}{\pi}$

Ans

22.   Let f be a differentiable function with $f(3) = 4$ and $f'(3) = 8$, and let g be the function defined by $g(x) = x\sqrt{f(x)}$. Which of the following is an equation of the line tangent to the graph of g at the point where $x = 3$?

(A) $y - 4 = 8(x - 3)$

(B) $y - 3 = 8(x - 6)$

(C) $y - 6 = 8(x - 3)$

(D) $y - 6 = 12(x - 3)$

(E) $y - 6 = -14(x - 3)$

Ans

23.   Let f be the function defined by $f(x) = x^{2/3}(5 - 2x)$. f is increasing on the interval

(A) $x < -\dfrac{5}{2}$ (B) $x > 0$ (C) $x < 1$ (D) $0 < x < \dfrac{5}{8}$ (E) $0 < x < 1$

Ans

24. Let R be the region in the first quadrant bounded by the x-axis and the curve $y = 2x - x^2$. The volume produced when R is revolved about the x-axis is

(A) $\dfrac{16\pi}{15}$ (B) $\dfrac{8\pi}{3}$ (C) $\dfrac{4\pi}{3}$ (D) 16π (E) 8π

Ans

25. What are all values of k for which the graph of $y = 2x^3 + 3x^2 + k$ will have three distinct x-intercepts?

(A) all $k < 0$

(B) all $k > -1$

(C) all k

(D) $-1 < k < 0$

(E) $0 < k < 1$

Ans

26. Use the Trapezoid Rule with $n = 4$ to approximate the integral $\displaystyle\int_1^5 f(x)\, dx$ for the function f whose graph is shown at the right.

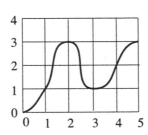

graph of $y = f(x)$

(A) 7

(B) 8

(C) 9

(D) 10

(E) 11

Ans

27. A point moves so that x, its distance from the origin at time t, $t \geq 0$ is given by:
$x(t) = \cos^3 t$. The first time interval in which the point is moving to the right is

(A) $0 < t < \frac{\pi}{2}$

(B) $\frac{\pi}{2} < t < \pi$

(C) $\pi < t < \frac{3\pi}{2}$

(D) $\frac{3\pi}{2} < t < 2\pi$

(E) none of these

Ans

28.   Which of the following is the solution to the differential equation $\frac{dy}{dx} = \frac{x}{y}$, where $y(-2) = -1$?

(A) $y = \sqrt{x^2 - 3}$ for $-\sqrt{3} < x < \sqrt{3}$

(B) $y = -\sqrt{x^2 - 3}$ for $x > \sqrt{3}$

(C) $y = \sqrt{x^2 - 3}$ for $x > \sqrt{3}$

(D) $y = \sqrt{x^2 - 3}$ for $x < -\sqrt{3}$

(E) $y = -\sqrt{x^2 - 3}$ for $x < -\sqrt{3}$

Ans

EXAM III
CALCULUS AB
SECTION I PART B
Time–50 minutes
Number of questions–17

**A GRAPHING CALCULATOR IS REQUIRED FOR SOME QUESTIONS ON
THIS PART OF THE EXAMINATION**

Directions: Solve each of the following problems, using the available space for scratchwork. After examining the form of the choices, decide which is the best of the choices given and fill in the box. Do not spend too much time on any one problem.

In this test:

(1)   The <u>exact</u> numerical value of the correct answer does not always appear among the choices given. When this happens, select from among the choices the number that best approximates the exact numerical value.

(2)   Unless otherwise specified, the domain of a function f is assumed to be the set of all real numbers x for which $f(x)$ is a real number.

(3)   The inverse of a trigonometric function f may be indicated using the inverse function notation f^{-1} or with the prefix "arc" (e.g., $\sin^{-1}x = \arcsin x$).

---

1.   The derivative of the function g is $g'(x) = \cos(\sin x)$. At the point where $x = 0$ the graph of g

    I. is increasing,   II. is concave down,   III. attains a relative maximum point.

    (A) I only   (B) II only   (C) III only   (D) I and III only   (E) I, II, III

Ans
⬚

---

2.   The approximate *average* rate of change of the function $f(x) = \int_{0}^{x} \sin(t^2)\, dt$ over the interval $[1, 3]$ is

    (A) 0.19   (B) 0.23   (C) 0.27   (D) 0.31   (E) 0.35

Ans
⬚

3.   When $\int_{-1}^{5} \sqrt{x^3 - x + 1} \, dx$ is approximated by using the mid-points of 3 rectangles of equal width, then the approximation is nearest to

(A) 22.6   (B) 22.9   (C) 23.2   (D) 23.5   (E) 23.8

Ans

☐

4.   Find the total area of the regions between the graph of the curve $y = x^3 - 5x^2 + 4x$ and the x-axis.

(A) 11.74

(B) 11.77

(C) 11.80

(D) 11.83

(E) 11.86

Ans

☐

5.   The graph of $y = \dfrac{\sin x}{x}$ has

I.   a vertical asymptote at $x = 0$

II. a horizontal asymptote at $y = 0$

III. an infinite number of zeros

(A) I only

(B) II only

(C) III only

(D) I and III only

(E) II and III only

Ans

☐

6. The graph of the function f is shown at the right. The graphs of the five functions:

$y = f(x+1),$

$y = f(x)+1,$

$y = f(-x),$

$y = f'(x)$ and

$y = \int_1^x f(t)\, dt$

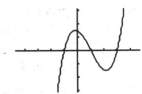

the graph of f

are shown in the *wrong* order.

The correct order is

(A) II, IV, III, V, I

(B) IV, II, III, I, V

(C) IV, II, III, V, I

(D) IV, III, II, V, I

(E) II, IV, III, I, V

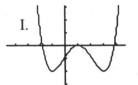

I.

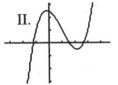

II.

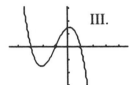

III.

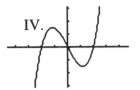

IV.

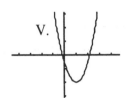

V.

Ans

7. The region in the first quadrant bounded above by the graph of $y = \sqrt{x}$ and below by the x-axis on the interval $[0, 4]$ is revolved about the x-axis. If a plane perpendicular to the x-axis at the point where $x = k$ divides the solid into parts of equal volume, then $k =$

(A) 2.77   (B) 2.80   (C) 2.83   (D) 2.86   (E) 2.89

Ans

8. The graph of the **derivative** of a function f is shown to the right. If the graph of f' has horizontal tangents at $x = -2$ and 1, which of the following is true about the function f?

graph of the derivative of f

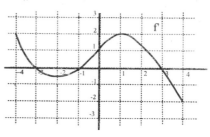

I. f is increasing on the interval $(-2, 1)$.

II. f is continuous at $x = 0$.

III. The graph of f has an inflection point at $x = -2$.

(A) I only   (B) II only   (C) III only   (D) II and III only   (E) I, II, III

Ans

☐

9. The area of the region completely bounded by the curve $y = -x^2 + 2x + 4$ and the line $y = 1$ is

(A) 8.7

(B) 9.7

(C) 10.7

(D) 11.7

(E) 12.7

Ans

☐

10. If functions f and g are defined so that $f'(x) = g'(x)$ for all real numbers x with $f(1) = 2$ and $g(1) = 3$, then the graph of f and the graph of g

(A) intersect exactly once;

(B) intersect no more than once;

(C) do not intersect;

(D) could intersect more than once;

(E) have a common tangent at each point of tangency.

Ans

☐

11. The graph of a function f whose domain
     is the interval $[-4, 4]$ is shown in the
     figure. If the graph of f has horizontal
     tangents at $x = -1.5$ and 2, which of the
     following statements are true?

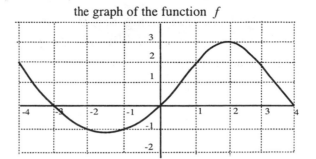

the graph of the function f

   I. The average rate of change of f over the
       interval from $x = -2$ to $x = 3$ is $\frac{1}{5}$.

   II. The slope of the tangent line at the point
        where $x = 2$ is 0.

   III. The left-sum approximation of $\displaystyle\int_{-1}^{3} f(t)\, dt$ with 4 equal subdivisions is 4.

   A) I only   (B) I and II only   (C) II and III only   (D) I and III only   (E) I, II, III

Ans

[   ]

12. One ship traveling west is $W(t)$ nautical miles west of a lighthouse and a second ship
     traveling south is $S(t)$ nautical miles south of the lighthouse at time t (hours). The graphs
     of W and S are shown below. At what approximate rate is the distance between the ships
     increasing at $t = 1$? (nautical miles per hour = knots)

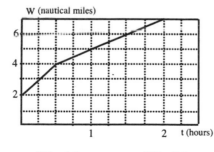

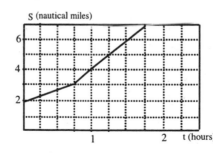

   (A) 1 knot   (B) 4 knots   (C) 7 knots   (D) 10 knots   (E) 13 knots

Ans

[   ]

13. Two particles move along the x-axis and their positions at time $0 \le t \le 2\pi$ are given by $x_1 = \cos 2t$ and $x_2 = e^{(t-3)/2} - 0.75$. For how many values of t do the two particles have the same velocity?

(A) 0

(B) 1

(C) 2

(D) 3

(E) 4

Ans

14. The line $x - 2y + 9 = 0$ is tangent to the graph of $y = f(x)$ at $(3, 6)$ and is also parallel to the line through $(1, f(1))$ and $(5, f(5))$. If f is differentiable on the closed interval $[1, 5]$ and $f(1) = 2$, find $f(5)$.

(A) 2

(B) 3

(C) 4

(D) 5

(E) none of these

Ans

15.   If $\dfrac{d}{dx}[f(x)] = g(x)$ and $\dfrac{d}{dx}[g(x)] = f(3x)$, then $\dfrac{d^2}{dx^2}[f(x^2)]$ is

(A) $4x^2 f(3x^2) + 2g(x^2)$

(B) $f(3x^2)$

C) $f(x^4)$

(D) $2xf(3x^2) + 2g(x^2)$

(E) $2xf(3x^2)$

Ans

16.   The point $(1, 9)$ lies on the graph of an equation $y = f(x)$ for which $\dfrac{dy}{dx} = 4x\sqrt{y}$

where $x \geq 0$ and $y \geq 0$. When $x = 0$ the value of y is

(A) 6

(B) 4

(C) 2

(D) $\sqrt{2}$

(E) 0

Ans

17.

x	-2	1	4	7
$g(x)$	3	1	5	-2

Let g be a continuous, differentiable function defined for all real numbers. Selected values of g are given in the table above. The graph of g must have

(A)   two points of inflection and at least one relative maximum

(B)   two zeros and at least one relative minimum

(C)   one zero and two points of inflection

(D)   one point of inflection and at least two relative maxima

(E)   at least one point of inflection and at least one relative minimum

Ans

`EXAM II
CALCULUS AB
SECTION II, PART A
Time–45 minutes
Number of problems–3

A graphing calculator is required for some problems or parts of problems.

- Before you begin Part A of Section II, you may wish to look over the problems before starting to work on them. It is not expected that everyone will be able to complete all parts of all problems and you will be able to come back to Part A (without a calculator), if you have time after Part B. All problems are given equal weight, but the parts of a particular solution are not necessarily given equal weight.

- You should write all work for each problem in the space provided. Be sure to write clearly and legibly. If you make an error, you may save time by crossing it out rather than trying to erase it. Erased or crossed out work will not be graded.

- SHOW ALL YOUR WORK. Clearly label any functions, graphs, tables, or other objects you use. You will be graded on the correctness and completeness of your methods as well as your final answers. Answers without supporting work may not receive credit.

- Justifications require that you give mathematical (noncalculator) reasons.

- You are permitted to use your calculator in Part A to solve an equation, find the derivative of a function at a point, or calculate the value of a definite integral. However, you must clearly indicate in your exam booklet the setup of your problem, namely the equation, function, or integral you are using. If you use other built-in features or programs, you must show the mathematical steps necessary to produce your results.

- Your work must be expressed in mathematical notation rather than calculator syntax. For example,
$$\int_{1}^{5} x^2 \, dx$$ may not be written as $\text{fnInt}(X^2, X, 1, 5)$.

- Unless otherwise specified, answers (numeric or algebraic) need not be simplified.

- If you use decimal approximations in your calculations, you will be graded on accuracy. Unless otherwise specified, your final answers should be accurate to three places after the decimal point.

- Unless otherwise specified, the domain of a function f is assumed to be the set of all real numbers x for which $f(x)$ is a real number.

THE EXAM BEGINS ON THE NEXT PAGE

PLEASE TURN OVER

1.   Let R be the region in the first quadrant bounded above by the graph of $f(x) = 3\cos x$ and below by the graph of $g(x) = e^{x^2}$.

   (a)   Find the area of region R.

   (b)   Set up, but <u>do</u> <u>not</u> <u>integrate</u> an integral expression in terms of a single variable for the volume of the solid generated when R is revolved about the <u>x-axis</u>.

   (c)   Let the base of a solid be the region R. If all cross sections perpendicular to the x-axis are squares, set up, but <u>do</u> <u>not</u> <u>integrate</u> an integral expression in terms of a single variable for the volume of the solid.

2.   Let f be the function defined by $f(x) = \ln(x + 1) - \sin^2 x$ for $0 \le x \le 3$.

(a) Find the x-intercepts of the graph of f.

(b) Find the intervals on which f is increasing.

(c) Find the absolute maximum and the absolute minimum value of f. Justify your answer.

3.　　　Let $f(x) = 4 - x^2$. For $0 < p < 2$, let $A(p)$ be the area of the triangle formed by the coordinate axes and the line tangent to the graph of f at the point $(p, 4 - p^2)$.

(a) Find $A(2)$.

(b) For what value of p is $A(p)$ a minimum?

A CALCULATOR MAY **NOT** BE USED ON THIS PART OF THE EXAMINATION.
DURING THE TIMED PORTION FOR PART B, YOU MAY GO BACK AND CONTINUE TO WORK
ON THE PROBLEMS IN PART A WITHOUT THE USE OF A CALCULATOR.

4.   Let $f(x) = x^3 + px^2 + qx$.

     (a) Find the values of p and q so that $f(-1) = -8$ and $f'(-1) = 12$.

     (b) Find the value of p so that the graph of f changes concavity at $x = 2$.

     (c) Under what conditions on p and q will the graph of f be increasing everywhere.

5.   A car is moving along a straight road from A to B, starting from A at time $t = 0$. Below is
      a graph of the car's velocity (positive direction from A to B), plotted against time.

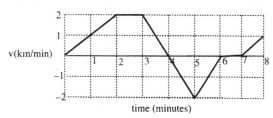

time (minutes)

(a) How many kilometers away from A is the car at time $t = 6$?

(b) At what time does the car change direction? Explain briefly.

(c) On the axes provided, sketch a graph of the acceleration of the car.

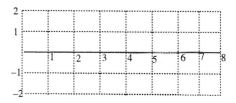

6.   Consider the curve given by the equation $y^3 - 3xy = 2$.

     (a) Find $\dfrac{dy}{dx}$.

     (b) Write an equation for the line tangent to the curve at the point $(1, 2)$. Use this tangent line to approximate y when $x = 1.3$.

     (c) Find $\dfrac{d^2y}{dx^2}$ at the point $(1, 2)$.

     (d) Is your approximation in Part (b) an overestimate or and underestimate. Justify your answer.

EXAM IV
CALCULUS AB
SECTION I PART A
Time–55 minutes
Number of questions–28

A CALCULATOR MAY NOT BE USED ON THIS PART OF THE EXAMINATION

Directions: Solve each of the following problems, using the available space for scratchwork. After examining the form of the choices, decide which is the best of the choices given and fill in the box. Do not spend too much time on any one problem.

In this test:

(1) Unless otherwise specified, the domain of a function f is assumed to be the set of all real numbers x for which $f(x)$ is a real number.

(2) The inverse of a trigonometric function f may be indicated using the inverse function notation f^{-1} or with the prefix "arc" (e.g., $\sin^{-1} x = \arcsin x$).

---

1. What is $\lim\limits_{x \to 0} \left(\dfrac{\dfrac{1}{x-1} + 1}{x} \right)$?

    (A) -1 (B) 0 (C) 1 (D) 2 (E) the limit does not exist

    Ans

---

2. $\displaystyle\int \frac{e^{\sqrt{x}}}{2\sqrt{x}}\, dx$

    (A) $\ln \sqrt{x} + C$ (B) $x + C$ (C) $e^x + C$ (D) $\frac{1}{2} e^{2\sqrt{x}} + C$ (E) $e^{\sqrt{x}} + C$

    Ans

---

3.   If $y = \dfrac{3}{4 + x^2}$, then $\dfrac{dy}{dx} =$

(A) $\dfrac{3}{2x}$

(B) $\dfrac{3x}{\left(1 + x^2\right)^2}$

(C) $\dfrac{6x}{\left(4 + x^2\right)^2}$

(D) $\dfrac{-6x}{\left(4 + x^2\right)^2}$

(E) $\dfrac{-3}{\left(4 + x^2\right)^2}$

Ans

☐

4.   If $F(x) = \displaystyle\int_1^x (\cos 6t + 1)\, dt$, then $F'(x) =$

(A) $\sin 6x + x$ (B) $\cos 6x + 1$ (C) $\dfrac{1}{6}\sin 6x + x$

(D) $-\dfrac{1}{6}\sin 6x + 1$ (E) $\sin 6x + 1$

Ans

☐

5.   Consider the curve $x + xy + 2y^2 = 6$. The slope of the line tangent to the curve at the point $(2,1)$ is

(A) $\dfrac{2}{3}$

(B) $\dfrac{1}{3}$

(C) $-\dfrac{1}{3}$

(D) $-\dfrac{1}{5}$

(E) $-\dfrac{3}{4}$

Ans

☐

6. $\lim\limits_{h \to 0} \dfrac{3\left(\frac{1}{2} + h\right)^5 - 3\left(\frac{1}{2}\right)^5}{h} =$

(A) 0

(B) 1

(C) $\dfrac{15}{16}$

(D) the limit does not exist

(E) the limit can not be determined

Ans

---

7.   If $p(x) = (x - 1)(x + k)$ and if the line tangent to the graph of p at the point $(4, p(4))$ is parallel to the line $5x - y + 6 = 0$, then $k =$

(A) 2

(B) 1

(C) 0

(D) –1

(E) –2

Ans

---

8.   If $\cos x = e^y$ and $0 < x < \dfrac{\pi}{2}$, what is $\dfrac{dy}{dx}$ in terms of x ?

(A) $- \tan x$

(B) $- \cot x$

(C) $\cot x$

(D) $\tan x$

(E) $\csc x$

Ans

9. At $t = 0$, a particle starts at the origin with a velocity of 6 feet per second and moves along the x-axis in such a way that at time t its acceleration is $12t^2$ feet per second per second. Through how many feet does the particle move during the first 2 seconds?

    (A) 16 ft

    (B) 20 ft

    (C) 24 ft

    (D) 28 ft

    (E) 32 ft

    Ans

10. When the area of an expanding square, in square units, is increasing three times as fast as its side is increasing, in linear units, the side is

    (A) $\frac{2}{3}$ (B) $\frac{3}{2}$ (C) 3   (D) 2   (E) 1

    Ans

11. The average (mean) value of $\frac{1}{x}$ over the interval $1 \leq x \leq e$ is

    (A) 1   (B) $\frac{1}{e}$ (C) $\frac{1}{e^2} - 1$ (D) $\frac{1+e}{2}$ (E) $\frac{1}{e-1}$

    Ans

12. What is $\lim\limits_{x\to\infty} \dfrac{3x^2+1}{(3-x)(3+x)}$?

(A) –9   (B) –3   (C) 1   (D) 3   (E) The limit does not exist.

Ans

☐

13. If $\displaystyle\int_{-2}^{2} (x^7+k)\,dx = 16,$ then $k =$

(A) –12

(B) 12

(C) –4

(D) 4

(E) 0

Ans

☐

14. Consider the function f defined on $\frac{\pi}{2} \le x \le \frac{3\pi}{2}$ by $f(x) = \dfrac{\tan x}{\sin x}$ for all $x \ne \pi$. If f is continuous at $x = \pi$, then $f(\pi) =$

(A) 2

(B) 1

(C) 0

(D) –1

(E) –2

Ans

☐

15. The function $f(x) = x^4 - 18x^2$ has a relative minimum at $x =$

(A) 0 and 3 only

(B) 0 and −3 only

(C) −3 and 3 only

(D) 0 only

(E) −3, 0, 3

Ans

16. The graph of $y = 3x^5 - 10x^4$ has an inflection point at

(A) $(0, 0)$ and $(2, -64)$

(B) $(0, 0)$ and $(3, -81)$

(C) $(0, 0)$ only

(D) $(-3, 81)$ only

(E) $(2, -64)$ only

Ans

17. The composite function h is defined by $h(x) = f[g(x)]$, where f and g are functions whose graphs are shown below.

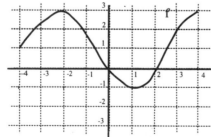 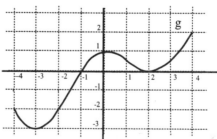

The number of points on the graph of h where there are horizontal tangent lines is

(A) 3   (B) 4   (C) 5   (D) 6   (E) 7

Ans

9) $x + xy + 2y^2 = 6$

$\frac{d}{dx}x + \frac{d}{dx}(xy) + \frac{d}{dx}(2y^2) = \frac{d}{dx}6$

$1 + (xy' + y) + 2y'(2y) = 0$

$1 + xy' + y + 4yy' = 0$

$xy' + 4yy' = -(y+1)$

$(x + 4y)\,y' = \dfrac{-y+1}{x+4y}$

$= \dfrac{-1+1}{2+4}$

$$x^3 - 5x^2 + 4x$$

$$\int_0^1 x^3 - 5x^2 + 4x$$

$$\int_1^4 x^3 - 5x^2 + 4x$$

18. The region in the first quadrant bounded by the graph of $y = \text{Arcsin } x$, $y = \dfrac{\pi}{2}$ and the y-axis, is rotated about the y-axis. The volume of the solid generated is given by

(A) $\pi \displaystyle\int_0^{\pi/2} y^2 \, dy$
(B) $\pi \displaystyle\int_0^1 (\text{Arcsin } x)^2 \, dx$
(C) $\pi \displaystyle\int_0^{\pi/2} (\text{Arcsin } x)^2 \, dx$

(D) $\pi \displaystyle\int_0^{\pi/2} (\sin y)^2 \, dy$
(E) $\pi \displaystyle\int_0^1 (\sin y)^2 \, dy$

Ans

19. Find the coordinates of the absolute maximum point for the curve $y = xe^{-kx}$ where k is a fixed positive number.

(A) $\left(\dfrac{1}{k}, \dfrac{1}{ke}\right)$
(B) $\left(\dfrac{-1}{k}, \dfrac{-e}{k}\right)$
(C) $\left(\dfrac{1}{k}, \dfrac{1}{e^k}\right)$
(D) $(0, 0)$
(E) there is no maximum

Ans

20. The slope field for a differential equation $\dfrac{dy}{dx} = f(x, y)$ is given in the figure. The slope field corresponds to which of the following differential equations?

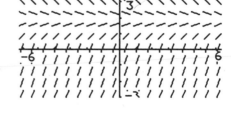

(A) $\dfrac{dy}{dx} = 2 - \ln x$

(B) $\dfrac{dy}{dx} = 2 - e^{-x}$

(C) $\dfrac{dy}{dx} = y - 2y^2$

(D) $\dfrac{dy}{dx} = 2 - y$

(E) $\dfrac{dy}{dx} = -x^2$

Ans

21. If y is a function of x such that $y' > 0$ for all x and $y'' < 0$ for all x, which of the following could be part of the graph of $y = f(x)$?

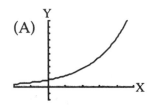 (A)   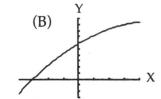 (B)   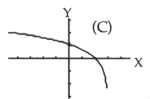 (C)

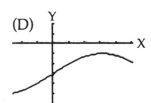 (D)   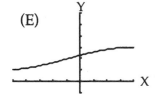 (E)

Ans

☐

22. Use the Trapezoid Rule with $n = 3$ to approximate the area under $y = x^2$ from $x = 1$ to $x = 4$.

(A) $\dfrac{45}{3}$

(B) $\dfrac{43}{3}$

(C) $\dfrac{43}{2}$

(D) 43

(E) 21

Ans

☐

23. If $f(x) = \dfrac{x^2+1}{e^x}$, then the graph of f is decreasing and concave down on the interval

(A) $(-\infty, 0)$ (B) $(0, 1)$ (C) $(1, 3)$ (D) $(3, 4)$ (E) $(4, \infty)$

Ans

☐

24. The number of bacteria in a culture is growing at a rate of $1500e^{3t/4}$ per unit of time t.
     At $t = 0$, the number of bacteria present was 2,000. Find the number present at $t = 4$.

(A) $2000\,e^3$

(B) $6000\,e^3$

(C) $2000\,e^6$

(D) $1500\,e^6$

(E) $1500\,e^3 + 500$

Ans

25. A region in the plane is bounded by the graph of $y = \dfrac{1}{x}$, the x-axis, the line $x = m$ and
     the line $x = 3m$, $m > 0$. The area of this region

(A) is independent of m

(B) increases as m increases

(C) decreases as m increases

(D) decreases for all $m < \dfrac{1}{3}$

(E) increases for all $m < \dfrac{1}{3}$

Ans

26. The formula $x(t) = \ln t + \dfrac{t^2}{18} + 1$ gives the position of an object moving along the x-axis during the time interval $1 \le t \le 5$. At the instant when the acceleration of the object is zero, the velocity is

(A) 0   (B) $\dfrac{1}{3}$ (C) $\dfrac{2}{3}$ (D) 1   (E) undefined

Ans

27. $\displaystyle\int 6 \sin x \cos^2 x \, dx \ =$

(A) $2 \sin^3 x + C$

(B) $-2 \sin^3 x + C$

(C) $2 \cos^3 x + C$

(D) $-2 \cos^3 x + C$

(E) $3 \sin^2 x \cos^2 x + C$

Ans

28. If for all $x > 0$, $G(x) = \displaystyle\int_1^x \sin(\ln 2t) \, dt$, then the value of $G''\!\left(\dfrac{1}{2}\right)$ is

(A) 0

(B) $\dfrac{1}{2}$

(C) 1

(D) 2

(E) undefined

Ans

EXAM IV
CALCULUS AB
SECTION I PART B
Time–50 minutes
Number of questions–17

A GRAPHING CALCULATOR IS REQUIRED FOR SOME QUESTIONS ON
THIS PART OF THE EXAMINATION

Directions: Solve each of the following problems, using the available space for scratchwork. After examining the form of the choices, decide which is the best of the choices given and fill in the box. Do not spend too much time on any one problem.

In this test:

(1) The <u>exact</u> numerical value of the correct answer does not always appear among the choices given. When this happens, select from among the choices the number that best approximates the exact numerical value.

(2) Unless otherwise specified, the domain of a function f is assumed to be the set of all real numbers x for which $f(x)$ is a real number.

(3) The inverse of a trigonometric function f may be indicated using the inverse function notation f^{-1} or with the prefix "arc" (e.g., $\sin^{-1} x = \arcsin x$).

---

1. The graph of the **second derivative** of a function f is shown at the right. Which of the following is true?

    I. The graph of f has an inflection point at $x = -1$.
    II. The graph of f is concave down on the interval $(-1, 3)$.
    III. The graph of the derivative function f' is increasing at $x = 1$.

    the graph of f''

    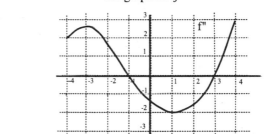

    (A) I only   (B) II only   (C) III only   (D) I and II only   (E) I, II, III

    Ans
    ☐

---

2. If the function f is continuous for all positive real numbers and if $f(x) = \dfrac{\ln x^2 - x \ln x}{x - 2}$ when $x \neq 2$, then $f(2) =$

    (A) -1 (B) -2 (C) $-e$ (D) $-\ln 2$ (E) undefined

    Ans
    ☐

---

3. The graph of the function f is shown at the right. At which point on the graph of f are all the following true?

$f(x) > 0$, and $f'(x) < 0$ and $f''(x) < 0$

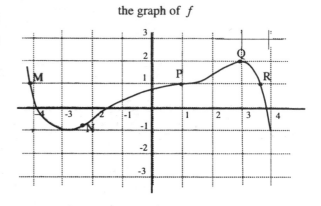
the graph of f

(A) M (B) N (C) P (D) Q (E) R

Ans

4. When using the substitution $u = \sqrt{1 + x}$, an antiderivative of $f(x) = 60x\sqrt{1+x}$ is

(A) $20u^3 - 60u + C$

(B) $15u^4 - 30u^2 + C$

(C) $30u^4 - 60u^2 + C$

(D) $24u^5 - 40u^3 + C$

(E) $12u^6 - 20u^4 + C$

Ans

5. At $x = 0$, which of the following statements is TRUE of the function f defined by
    $f(x) = \sqrt{x^2 + .0001}$.

I. f is discontinuous   II. f has a horizontal tangent   III. f' is undefined

(A) I only   (B) II only   (C) III only   (D) I and III only   (E) I, II, III

Ans

6.   Functions f and g are defined by $f(x) = \dfrac{1}{x^2}$ and $g(x) = \arctan x$. What is the approximate

value of x for which $f'(x) = g'(x)$?

(A) -3.36 (B) -2.86 (C) -2.36 (D) 1.36 (E) 2.36

Ans

☐

.7.   The area of the region bounded below by $f(x) = x^2 - 7x + 10$ and above by
$g(x) = \ln(x - 1)$ is closest to

(A) 7.35

(B) 7.36

(C) 7.38

(D) 7.40

(E) 7.42

Ans

☐

8.   The average rate of change of the function $f(x) = \displaystyle\int_0^x \sqrt{1 + \cos(t^2)}\ dt$ over the

interval $[1, 3]$ is nearest to

(A) 0.85

(B) 0.86

(C) 0.87

(D) 0.88

(E) 0.89

Ans

☐

9.   The graph of the *derivative* of f is shown at the
      right. If the graph of f' has horizontal tangents
      at $x = -2$ and 0, which of the following is true
      about the function f?

      graph of the derivative of f

      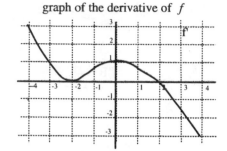

      I. f is decreasing at $x = 0$.

      II. f has a local maximum at $x = 2$.

      III. The graph of f is concave up at $x = -1$.

      (A) I only   (B) II only   (C) I and II only   (D) II and III only   (E) I, II, III

                                                                                    Ans

10.   Let f be the function defined by $f(x) = 3 + \int_2^x \frac{20}{1+t^2}\,dt$. Which of the following is an
      equation of the line tangent to the graph of f at the point where $x = 2$?

      (A) $y = 4(x - 2)$

      (B) $y - 3 = 7(x - 2)$

      (C) $y - 2 = 4(x - 3)$

      (D) $y - 3 = 4(x - 2)$

      (E) $y - 2 = 7(x - 3)$

                                                                                    Ans

11.   If a left Riemann sum overapproximates the definite integral $\int_0^4 f(x)\,dx$ and a trapezoid sum

      underapproximates the integral $\int_0^4 f(x)\,dx$, which of the following could be a graph of $y = f(x)$?

      (A)   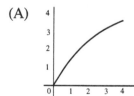   (B)   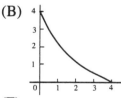   (C)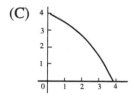

      (D)   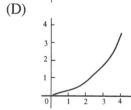   (E)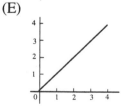

                                                                                    Ans

12. The function V whose graph is sketched below gives the volume of air, $V(t)$, (measured in cubic inches) that a man has blown into a balloon after t seconds.

$$\left(V = \frac{4}{3}\pi r^3 \right)$$

The rate at which the radius is changing after 6 seconds is nearest to

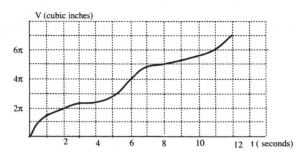

(A) 0.05 in/sec   (B) 0.12 in/sec   (C) 0.21 in/sec   (D) 0.29 in/sec   (E) 0.37 in/sec

Ans

---

13. At how many points on the inteval $-2\pi \le x \le 2\pi$ does the tangent to the graph of the curve $y = x\cos x$ have slope $\frac{\pi}{2}$?

(A) 5

(B) 4

(C) 3

(D) 2

(E) 1

Ans

14.   The graph of $y = f'(x)$, the derivative of a function f, is
      a line and a quarter-circle shown in the diagram. If
      $f(2) = 3$, then $f(6) =$

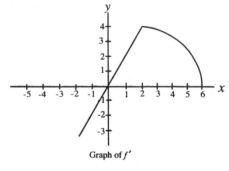

Graph of f'

(A) 4   (B) 7   (C) 3+4π   (D) 7+4π   (E) 11

Ans

15.   Let the base of a solid be the first quadrant region bounded above by the graph of $y = \sqrt{x}$ and
      below by the x-axis on the interval $[0, 4]$. If every cross section perpendicular to the x-axis is an
      isosceles right triangle with one leg in the base, then the volume of the solid is

      (A) 2

      (B) 4

      (C) 8

      (D) 16

      (E) 32

Ans

16. Let the function F be defined on the interval $[0, 8]$ by $F(x) = \int_0^x f(t)\, dt$, where the

   graph of f is shown below. The graph of f consists of four line segments and a semicircle.

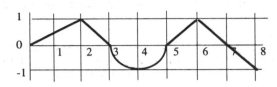

graph of $y = f(t)$

   In which of the following intervals does F have a zero?

   I. $4 < x < 5$ II. $5 < x < 6$ III. $6 < x < 7$

   (A) I only   (B) II only   (C) III only   (D) I and II only   (E) I and III only

Ans

17. The change in N, the number of bacteria in a culture dish at time t, is given by: $\dfrac{dN}{dt} = 2N$.

   If $N = 3$ when $t = 0$, the approximate value of t when $N = 1210$ is

   (A) 2   (B) 3   (C) 4   (D) 5   (E) 6

Ans

EXAM IV
CALCULUS AB
SECTION II, PART A
Time–45 minutes
Number of problems–3

A graphing calculator is required for some problems or parts of problems.

- Before you begin Part A of Section II, you may wish to look over the problems before starting to work on them. It is not expected that everyone will be able to complete all parts of all problems and you will be able to come back to Part A (without a calculator), if you have time after Part B. All problems are given equal weight, but the parts of a particular solution are not necessarily given equal weight.

- You should write all work for each problem in the space provided. Be sure to write clearly and legibly. If you make an error, you may save time by crossing it out rather than trying to erase it. Erased or crossed out work will not be graded.

- SHOW ALL YOUR WORK. Clearly label any functions, graphs, tables, or other objects you use. You will be graded on the correctness and completeness of your methods as well as your final answers. Answers without supporting work may not receive credit.

- Justifications require that you give mathematical (noncalculator) reasons.

- You are permitted to use your calculator in Part A to solve an equation, find the derivative of a function at a point, or calculate the value of a definite integral. However, you must clearly indicate in your exam booklet the setup of your problem, namely the equation, function, or integral you are using. If you use other built-in features or programs, you must show the mathematical steps necessary to produce your results.

- Your work must be expressed in mathematical notation rather than calculator syntax. For example,

$$\int_1^5 x^2 \, dx \quad \text{may not be written as} \quad \text{fnInt}(X^2, X, 1, 5).$$

- Unless otherwise specified, answers (numeric or algebraic) need not be simplified.

- If you use decimal approximations in your calculations, you will be graded on accuracy. Unless otherwise specified, your final answers should be accurate to three places after the decimal point.

- Unless otherwise specified, the domain of a function f is assumed to be the set of all real numbers x for which $f(x)$ is a real number.

THE EXAM BEGINS ON THE NEXT PAGE

PLEASE TURN OVER

1.   Let R be the region in the first quadrant bounded by the graph of $y = \frac{1}{x+1}$ and the line
      $x = 4$.

   (a) Use a left-hand Riemann sum with 4 subdivisions of equal length to approximate the
       area of region R.

   (b) Find the exact area of R.

   (c) Let $x = k$ be a vertical line dividing R into two regions of equal area. Find the value
       of k.

   (d) Find the volume of the solid with base R and whose cross sections cut by planes
       perpendicular to the x-axis are squares.

2. Two points, A and B, are located 275 ft apart on a level field At a given instant, a balloon
    is released at B and rises vertically at a constant rate of 2.5 ft/sec, and, at the same instant, a
    cat starts running from A to B at a constant rate of 5 ft/sec.

    (a) After 40 seconds, is the distance between the cat and the balloon decreasing or
         increasing? At what rate?

    (b) Describe what is happening to the distance between the cat and the balloon at
         $t = 50$ seconds.

    (c) Is there a time when the cat is closest to the balloon? If yes, find this time. If no,
         explain why?

3. Let $y(t)$ be the temperature, in degrees Fahrenheit, of a cup of tea at time t minutes, $t \geq 0$. Room temperature is $70°$ and the initial temperature of the tea is $180°$. The tea's temperature at time t is described by the differential equation $\dfrac{dy}{dt} = -0.1(y - 70)$, with the initial condition $y(0) = 180$.

(a) Use separation of variables to find an expression for y in terms of t, where t is measured in minutes.

(b) How hot is the tea after 10 minutes?

(c) If the tea is safe to drink when its temperature is less than $120°$, at what time is the tea safe to drink?

A CALCULATOR MAY **NOT** BE USED ON THIS PART OF THE EXAMINATION.
DURING THE TIMED PORTION FOR PART B, YOU MAY GO BACK AND CONTINUE TO WORK
ON THE PROBLEMS IN PART A WITHOUT THE USE OF A CALCULATOR.

4.   Let f be the function defined by $f(x) = x - 2\cos x$ on the closed interval $[0, 2\pi]$

(a) Showing your reasoning, determine the value of x at which f has its

  (i)   absolute maximum

  (ii)   absolute minimum

(b)   For what values of x is the graph of f concave down. Justify your answer.

(c)   Find the average value of f over the interval $[0, 2\pi]$

5.   Let g be the function given by $g(x) = \dfrac{x \cdot |x|}{x^2 + 1}$.

   (a) Determine whether the *derivative* of the function g is even, odd, or neither.

   (b) Find $g'(2)$.

   (c) Evaluate $\displaystyle\int_0^1 g(x)\, dx$.

   (d) Determine $\displaystyle\lim_{x \to \infty} g(x)$.

   (e) Find the range of g. Justify your answer.

6. The graph of a differentiable function f on the closed interval $[-4, 4]$ is shown at the right. The graph of f has horizontal tangents at $x = -3, -1$ and 2.

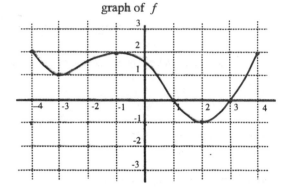

graph of f

Let $G(x) = \displaystyle\int_{-4}^{x} f(t)\, dt$ for $-4 \le x \le 4$.

(a) Find $G(-4)$.

(b) Find $G'(-1)$.

(c) On which interval or intervals is the graph of G concave down. Justify your answer.

(d) Find the value of x at which G has its maximum on the closed interval $[-4, 4]$. Justify your answer.

EXAM V
CALCULUS AB
SECTION I PART A
Time–55 minutes
Number of questions–28

A CALCULATOR MAY NOT BE USED ON THIS PART OF THE EXAMINATION

Directions: Solve each of the following problems, using the available space for scratchwork. After examining the form of the choices, decide which is the best of the choices given and fill in the box. Do not spend too much time on any one problem.

In this test:

(1) Unless otherwise specified, the domain of a function f is assumed to be the set of all real numbers x for which $f(x)$ is a real number.

(2) The inverse of a trigonometric function f may be indicated using the inverse function notation f^{-1} or with the prefix "arc" (e.g., $\sin^{-1} x = \arcsin x$).

---

1. If $y = \cos^2(2x)$, then $\dfrac{dy}{dx} =$

   (A) $2\cos 2x \sin 2x$

   (B) $-4\sin 2x \cos 2x$

   (C) $2\cos 2x$

   (D) $-2\cos 2x$

   (E) $4\cos 2x$

   Ans

---

2. A slope field for a differential equation $\dfrac{dy}{dx} = f(x,y)$ is given in the figure at the right. Which of the following statements are true?

   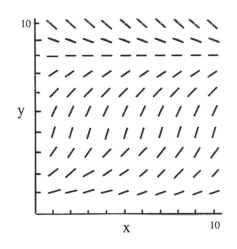

   I.   The value of $\dfrac{dy}{dx}$ at the point $(2, 2)$ is approximately 1.

   II. As y approaches 8 the rate of change of y approaches zero.

   III. All solution curves for the differential equation have the same slope for a given value of x.

   (A) I only   (B) II only   (C) I and II only   (D) II and III only   (E) I, II, III

   Ans

---

3. The slope of the line tangent to the graph of $y = \ln \sqrt{x}$ at $(e^2, 1)$ is

(A) $\dfrac{e^2}{2}$　　　　(B) $\dfrac{2}{e^2}$　　　　(C) $\dfrac{1}{2e^2}$　　　　(D) $\dfrac{1}{2e}$　　　　(E) $\dfrac{1}{e}$

Ans

4. Which of the following functions is both continuous and differentiable at all x in the interval $-2 \le x \le 2$?

(A) $f(x) = \left| x^2 - 1 \right|$

(B) $f(x) = \sqrt{x^2 - 1}$

(C) $f(x) = \sqrt{x^2 + 1}$

(D) $f(x) = \dfrac{1}{x^2 - 1}$

(E) none of these

Ans

5. Find the point on the graph of $y = \sqrt{x}$ between $(1, 1)$ and $(9, 3)$ at which the tangent to the graph has the same slope as the line through $(1, 1)$ and $(9, 3)$.

(A) $(1, 1)$

(B) $(2, \sqrt{2})$

(C) $(3, \sqrt{3})$

(D) $(4, 2)$

(E) none of the above

Ans

6. Consider the function $f(x) = \dfrac{x^4}{2} - \dfrac{x^5}{10}$. The *derivative* of f attains its maximum

    value at $x =$

    (A) 3

    (B) 4

    (C) 5

    (D) 0

    (E) there is no maximum

    Ans
    ☐

7. The acceleration, $a(t)$, of a body moving in a straight line is given in terms of time t by
    $a(t) = 4 - 6t$. If the velocity of the body is 20 at $t = 0$ and if $s(t)$ is the distance of the
    body from the origin at time t, what is $s(3) - s(1)$?

    (A) –10

    (B) 0

    (C) 10

    (D) 20

    (E) 30

    Ans
    ☐

8. $\lim\limits_{x \to 1} \left(\dfrac{\sqrt{x + 3} - 2}{1 - x} \right)$

    (A) 0.5

    (B) 0.25

    (C) 0

    (D) –0.25

    (E) –0.5

    Ans
    ☐

9.   Let f be defined by $f(x) = \begin{cases} \dfrac{x^2 - 2x + 1}{x - 1} & \text{for } x \neq 1 \\ k & \text{for} \quad x = 1. \end{cases}$

Determine the value of k for which f is continuous for all real x.

(A) 0

(B) 1

(C) 2

(D) 3

(E) none of the above

Ans

10.   The average value of $f(x) = e^{2x} + 1$ on the interval $0 \leq x \leq \frac{1}{2}$ is

(A) e (B) $\dfrac{e}{2}$ (C) $\dfrac{e}{4}$ (D) $2e - 1$ (E) $\dfrac{e^{2x} + 1}{2}$

Ans

11.   A particle moves along the x-axis in such a way that its velocity at time $t > 0$ is given by

$v = \dfrac{e^t}{t}$. At what value of t does v attain its minimum?

(A) 0

(B) 1

(C) e

(D) -1

(E) There is no minimum value of v.

Ans

12. $\int \dfrac{4x}{1 + x^2}\ dx =$

    (A) $4\operatorname{Arctan} x + C$ (B) $\dfrac{4}{x}\operatorname{Arctan} x + C$ (C) $\dfrac{1}{2}\ln(1 + x^2) + C$

    (D) $2\ln(1 + x^2) + C$ (E) $2x^2 + 4\ln|x| + C$

Ans ☐

13. Let $f(x) = x^4 + ax^2 + b$. The graph of f has a relative maximum at $(0, 1)$ and an inflection point when $x = 1$. The values of a and b are

    (A) $a = 1, \quad b = -6$

    (B) $a = 1, \quad b = 6$

    (C) $a = -6, \quad b = 5$

    (D) $a = -6, \quad b = 1$

    (E) $a = 6, \quad b = 1$

Ans ☐

14. $\displaystyle\int_{1}^{2} \dfrac{x^2 - x}{x^3}\ dx =$

    (A) $\ln 2 - \dfrac{1}{2}$ (B) $\ln 2 + \dfrac{1}{2}$ (C) $\dfrac{1}{2}$ (D) 0 (E) $\dfrac{1}{4}$

Ans ☐

15. The edge of a cube is increasing at the uniform rate of 0.2 inches per second. At the instant when the total surface area becomes 150 square inches, what is the rate of increase, in cubic inches per second, of the volume of the cube?

(A) 5 in^3/sec

(B) 10 in^3/sec

(C) 15 in^3/sec

(D) 20 in^3/sec

(E) 25 in^3/sec

Ans

16. $\displaystyle\int_0^{\sqrt{3}} \frac{x\,dx}{\sqrt{1+x^2}} =$

(A) $\dfrac{1}{2}$

(B) 1

(C) 2

(D) ln 2

(E) Arctan $2 - \dfrac{\pi}{4}$

Ans

17. Which of the following is true about the graph of $f(x) = \ln\left|x^2 - 4\right|$ in the interval $(-2, 2)$?

(A) f is increasing.

(B) f attains a relative minimum at $(0, 0)$.

(C) f has a range of all real numbers.

(D) f is concave down.

(E) f has an asymptote at $x = 0$.

Ans

18. If $g(x) = \text{Arcsin } 2x$, then $g'(x) =$

(A) $2\text{Arccos } 2x$ (B) $2 \csc 2x \cot 2x$ (C) $\dfrac{2}{1 + 4x^2}$

(D) $\dfrac{2}{\sqrt{4x^2 - 1}}$ (E) $\dfrac{2}{\sqrt{1 - 4x^2}}$

Ans

19. $\displaystyle\int x(x^2 - 1)^4 \; dx \;\; =$

(A) $\dfrac{1}{10}(x^2)(x^2 - 1)^5 + C$

(B) $\dfrac{1}{10}(x^2 - 1)^5 + C$

(C) $\dfrac{1}{5}(x^3 - x)^5 + C$

(D) $\dfrac{1}{5}(x^2 - 1)^5 + C$

(E) $\dfrac{1}{5}(x^2 - x)^5 + C$

Ans

20. If $y = e^{kx}$, then $\dfrac{d^5 y}{dx^5} =$

(A) $k^5 e^x$

(B) $k^5 e^{kx}$

(C) $5! \, e^{kx}$

(D) $5! \, e^x$

(E) $5 e^{kx}$

Ans

21. The graph of f is shown at the right. Which of the following statements are true?

the graph of f

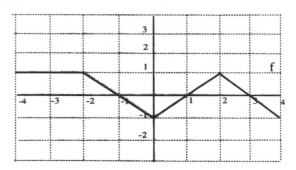

   I. $f(2) > f'(1)$

   II. $\displaystyle\int_0^1 f(x)\,dx > f'(3.5)$

   III. $\displaystyle\int_{-1}^1 f(x)\,dx > \int_{-1}^2 f(x)\,dx$

(A) I only   (B) II only   (C) I and II only   (D) II and III only   (E) I, II, III

Ans

22. If $g(x) = \sqrt{x}\,(x-1)^{2/3}$, then the domain of g' is

(A) $\left\{x \mid 0 < x\right\}$

(B) $\left\{x \mid x \neq 0 \text{ and } x \neq 1\right\}$

(C) $\left\{x \mid 0 < x < 1 \text{ or } x > 1\right\}$

(D) $\left\{x \mid 0 < x < 1\right\}$

(E) $\left\{x \mid \text{all real numbers}\right\}$

Ans

23. A particle moves along the x-axis so that its distance from the origin at time t is given by $10t - 4t^2$. What is the *total* distance covered by the point between $t = 1$ and $t = 2$?

(A) 1.0

(B) 1.5

(C) 2.0

(D) 2.5

(E) 3.0

Ans

24. At which point on the graph of $y = g(x)$ below is $g'(x) = 0$ and $g''(x) = 0$?

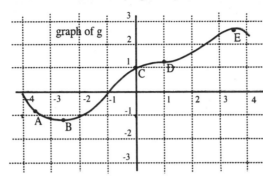

(A) A   (B) B   (C) C   (D) D   (E) E

Ans

25. If y is a differentiable function of x, then the slope of the tangent to the curve
     $xy - 2y + 4y^2 = 6$ at the point where $y = 1$ is

(A) $\frac{1}{12}$ (B) $-\frac{1}{10}$ (C) $-\frac{1}{6}$ (D) $\frac{1}{4}$ (E) $-\frac{5}{6}$

Ans

26. The area of the region bounded above by $y = 1 + \sec^2 x$, below by $y = 0$, on the left
     by $x = 0$ and on the right by $x = \frac{\pi}{4}$ is approximately

(A) 1   (B) 1.25   (C) 1.5   (D) 1.75   (E) 2

Ans

27. A solution of the equation $\dfrac{dy}{dx} + 2xy = 0$ that contains the point $(0, e)$ is

   (A) $y = e^{1-x^2}$

   (B) $y = e^{1+x^2}$

   (C) $y = e^{1-x}$

   (D) $y = e^{1+x}$

   (E) $y = e^{x^2}$

Ans

28. Which of the following are true about the function $F(x) = \displaystyle\int_{1}^{x} \ln(2t - 1)\, dt$?

   I. $F(1) = 0$ II. $F'(1) = 0$ $F''(1) = 1$

   (A) I and II only

   (B) I and III only

   (C) II and III only

   (D) I, II, III

   (E) none

Ans

EXAM V
CALCULUS AB
SECTION I   PART B
Time–50 minutes
Number of questions–17

A GRAPHING CALCULATOR IS REQUIRED FOR SOME QUESTIONS ON THIS PART OF THE EXAMINATION

Directions: Solve each of the following problems, using the available space for scratchwork. After examining the form of the choices, decide which is the best of the choices given and fill in the box. Do not spend too much time on any one problem.

In this test:

(1)   The exact numerical value of the correct answer does not always appear among the choices given. When this happens, select from among the choices the number that best approximates the exact numerical value.

(2)   Unless otherwise specified, the domain of a function f is assumed to be the set of all real numbers x for which $f(x)$ is a real number.

(3)   The inverse of a trigonometric function f may be indicated using the inverse function notation f^{-1} or with the prefix "arc" (e.g., $\sin^{-1} x = \arcsin x$).

---

1.   How many points of inflection does the graph of $y = \cos x + \dfrac{1}{3} \cos 3x - \dfrac{1}{5} \cos 5x$ have on the interval $0 \le x \le \pi$?

(A) 1

(B) 2

(C) 3

(D) 4

(E) 5

Ans

---

2.   Oil is leaking from a tanker at the rate of $R(t) = 500 e^{-0.2t}$ gallons per hour, where t is measured in hours. The amount of oil that has leaked out after 10 hours is closest to

(A) 2140 gals

(B) 2150 gals

(C) 2160 gals

(D) 2170 gals

(E) 2180 gals

Ans

---

3. The sale of lumber S (in millions of square feet) for the years 1980 to 1990 is modeled by the function

$$S(t) = 0.46\cos(0.45t + 3.15) + 3.4$$

where t is the time in years with $t = 0$ corresponding to the beginning of 1980. Determine the year when lumber sales were increasing at the greatest rate.

(A) 1982

(B) 1983

(C) 1984

(D) 1985

(E) 1986

Ans

4. The graph of f over the interval $[1, 9]$ is shown in the figure. Using the data in the figure, find a midpoint approximation with 4 equal subdivisions for $\int_{1}^{9} f(x)\, dx$.

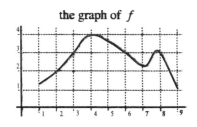

the graph of f

(A) 20   (B) 21   (C) 22   (D) 23   (E) 24

Ans

5. Let the base of a solid be the first quadrant region enclosed by the x-axis and one arch of the graph of $y = \sin x$. If all cross sections perpendicular to the x-axis are squares, then the volume of the solid is approximately

(A) 0.52

(B) 0.79

(C) 1.05

(D) 1.57

(E) 2.00

Ans

6. If $f(x) = 2x + \sin x$ and the function g is the inverse of f, then $g'(2) =$

(A) 0.32

(B) 0.34

(C) 0.36

(D) 0.38

(E) 0.40

Ans

☐

7. Administrators at Massachusetts General Hospital believe that the hospital's expenditures $E(B)$, measured in dollars, are a function of how many beds B are in use with
$$E(B) = 14000 + (B + 1)^2.$$

On the other hand, the number of beds B is a function of time t, measured in days, and it is estimated that
$$B(t) = 20 \sin\left(\frac{t}{10}\right) + 50.$$

At what rate are the expenditures decreasing when $t = 100$?

(A) 120 dollars/day

(B) 125 dollars/day

(C) 130 dollars/day

(D) 135 dollars/day

(E) 140 dollars/day

Ans

☐

8. Let f be a function that has domain $[-2, 5]$. The graph of f' is shown at the right. Which of the following statements are TRUE?

The graph of f'

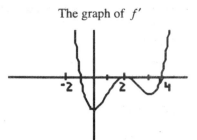

I. f has a relative maximum at $x = -1$.

II. f has an absolute minimum at $x = 0$.

III. The graph of f is concave down for $-2 < x < 0$.

IV. The graph of f has inflection points at $x = 0$ and $x = 2$ and $x = 3$.

(A) I, II, IV   (B) I, III, IV   (C) II, III, IV   (D) I, II, III   (E) I, II, III, IV

Ans

9. On which interval is the graph of $f(x) = 4x^{3/2} - 3x^2$ both concave down and increasing?

(A) $(0, 1)$

(B) $\left(0, \dfrac{1}{2}\right)$

(C) $\left(0, \dfrac{1}{4}\right)$

(D) $\left(\dfrac{1}{4}, \dfrac{1}{2}\right)$

(E) $\left(\dfrac{1}{4}, 1\right)$

Ans

10. The average rate of change of the function $f(x) = x^2 - \dfrac{1}{e^x}$ over the interval $[0, 3]$ equals the instantanous rate of change of f at $x =$

(A) 0.313   (B) 1.553   (C) 2.573   (D) 3.317   (E) 9.950

Ans

11. If $\sin 3x - 1 = \int\limits_{a}^{x} f(t)\, dt$, then the value of a is

(A) 0

(B) 1

(C) −1

(D) $\dfrac{\pi}{3}$

(E) $\dfrac{\pi}{6}$

Ans

12. If $xy^2 = 20$ and x is decreasing at the rate of 3 units per second, the rate at which y is changing when $y = 2$ is nearest to

(A) −0.6 units/sec

(B) −0.2 units/sec

(C) 0.2 units/sec

(D) 0.6 units/sec

(E) 1.0 units/sec

Ans

13. An approximation for $\int\limits_{-1}^{2} e^{\sin(1.5x-1)}\, dx$ using a right-hand Riemann sum with three equal subdivisions is nearest to

(A) 2.5

(B) 3.5

(C) 4.5

(D) 5.5

(E) 6.5

Ans

14. If $f(x)$ is defined on $-\pi \le x \le \pi$ and $\dfrac{dy}{dx} = \dfrac{\cos x}{x^2 + 1}$, which of the following statements about the graph of $y = f(x)$ is true?

   (A) The graph has no relative extremum.

   (B) The graph has one point of inflection and two relative extrema.

   (C) The graph has two points of inflection and one relative extremum.

   (D) The graph has two points of inflection and two relative extrema.

   (E) The graph has three points of inflection and two relative extrema.

Ans

15. The graph of the function f is shown at the right. If the function G is defined by

   $$G(x) = \int_{-4}^{x} f(t)\, dt, \quad \text{for } -4 \le x \le 4, \text{ which of}$$

   the following statements about G are true?

   graph of the function f

   I. G is increasing on $(1, 2)$.

   II. G is decreasing on $(-4, -3)$.

   III. $G(0) < 0$.

   (A) None   (B) II only   (C) III only   (D) II and III only   (E) I and II only

Ans

16. The function f is defined on all the reals such that $f(x) = \begin{cases} x^2 + kx - 3 & \text{for } x \le 1 \\ 3x + b & \text{for } x > 1. \end{cases}$

For which of the following values of k and b will the function f be both continuous and differentiable on its entire domain?

(A) $k = -1, b = -3$

(B) $k = 1, b = 3$

(C) $k = 1, b = 4$

(D) $k = 1, b = -4$

(E) $k = -1, b = 6$

Ans

17. A particle moves along the x-axis with velocity at time t given by: $v(t) = t + 2 \sin t$. If the particle is at the origin when $t = 0$, its position at the time when $v = 6$ is $x =$

(A) 17.159   (B) 19.159   (C) 23.141   (D) 29.201   (E) 39.309

Ans

EXAM V
CALCULUS AB
SECTION II, PART A
Time–45 minutes
Number of problems–3

A graphing calculator is required for some problems or parts of problems.

- Before you begin Part A of Section II, you may wish to look over the problems before starting to work on them. It is not expected that everyone will be able to complete all parts of all problems and you will be able to come back to Part A (without a calculator), if you have time after Part B. All problems are given equal weight, but the parts of a particular solution are not necessarily given equal weight.

- You should write all work for each problem in the space provided. Be sure to write clearly and legibly. If you make an error, you may save time by crossing it out rather than trying to erase it. Erased or crossed out work will not be graded.

- SHOW ALL YOUR WORK. Clearly label any functions, graphs, tables, or other objects you use. You will be graded on the correctness and completeness of your methods as well as your final answers. Answers without supporting work may not receive credit.

- Justifications require that you give mathematical (noncalculator) reasons.

- You are permitted to use your calculator in Part A to solve an equation, find the derivative of a function at a point, or calculate the value of a definite integral. However, you must clearly indicate in your exam booklet the setup of your problem, namely the equation, function, or integral you are using. If you use other built-in features or programs, you must show the mathematical steps necessary to produce your results.

- Your work must be expressed in mathematical notation rather than calculator syntax. For example, $\int_{1}^{5} x^2\, dx$ may not be written as $\text{fnInt}(X^2, X, 1, 5)$.

- Unless otherwise specified, answers (numeric or algebraic) need not be simplified.

- If you use decimal approximations in your calculations, you will be graded on accuracy. Unless otherwise specified, your final answers should be accurate to three places after the decimal point.

- Unless otherwise specified, the domain of a function f is assumed to be the set of all real numbers x for which $f(x)$ is a real number.

THE EXAM BEGINS ON THE NEXT PAGE

PLEASE TURN OVER

1.   The position of a particle moving on the x-axis at time $t > 0$ seconds is: $x(t) = e^t - \sqrt{t}$ feet.

(a) Find the average velocity of the particle over the interval $1 \le t \le 3$.

(b) In what direction and how fast is the particle moving at $t = 1$ seconds?

(c) For what values of t is the particle moving to the right?

(d) Find the position of the particle when its velocity is zero.

2. Water flowed into a tank at an increasing rate $r(t)$ from $t = 0$ to $t = 5$ minutes. The rate of flow, $r(t)$, in cubic meters per minute (m^3/min), was measured at one minute intervals with the result shown in the table below.

t	0	1	2	3	4	5
$r(t)$	4	5	7	11	12	14

(a) Give the best upper and lower estimates for the total amount of water that flowed into the tank for $0 \le t \le 5$. Indicate units of measure.

(b) Suppose you use the average of the upper and lower estimates found in part (a) as your approximation for the total amount of water that flowed into the tank, what is the maximum error for this approximation?

(c) You are now informed that for $1 \le t \le 3$ the rate of flow was exactly $r(t) = t^2 - t + 5$ m^3/min. What is the exact amount of water that flowed into the tank from $t = 1$ to $t = 3$?

3.   Let R be the first quadrant region enclosed by the graph of $y = 2e^{-x}$ and the line $x = k$.

   (a) Find the area of R in terms of k.

   (b) Find the volume of the solid generated when R is rotated about the x-axis in terms of k.

   (c) What is the volume in part (b) as $k \to \infty$?

A CALCULATOR MAY **NOT** BE USED ON THIS PART OF THE EXAMINATION.
DURING THE TIMED PORTION FOR PART B, YOU MAY GO BACK AND CONTINUE TO WORK
ON THE PROBLEMS IN PART A WITHOUT THE USE OF Λ CALCULATOR.

4.   Consider the differential equation $\dfrac{dy}{dx} = \dfrac{xy}{(x^2+4)}$.

(a) On the axes provided, sketch a slope field for the given differential equation at the fourteen points indicated.

(b) Sketch the solution curve that contains the point (–2, 2).

(c) Find a general solution to the differential equatiion.

(d) Find the particular solution to the differential equation that satisfies the initial condition $y(0) = 4$.

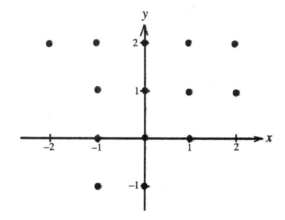

5.   Let f be the function defined by $f(x) = \dfrac{ax}{x^2 + 1}$, where a is a positive constant.

     (a) Find $\lim\limits_{x \to -\infty} f(x)$ and $\lim\limits_{x \to +\infty} f(x)$.

     (b) Write an expression for $f'(x)$ and use it to find the relative maximum and minimum value of f in terms of a. Justify your answers.

     (c) Find the value of a such that the average value of f over the interval $\left[0, \sqrt{2}\right]$ is $\dfrac{1}{2}$.

6. Let f be the function defined by $f(x) = \ln\left(\dfrac{x}{x+1}\right)$.

   (a) What is the domain of f?

   (b) Find $f'(x)$.

   (c) Find an equation for the tangent line to the graph of f at the point $(1, f(1))$.

   (d) Write an expression for $g'(x)$, where g is the inverse function of f.

EXAM VI
CALCULUS AB
SECTION I   PART A
Time–55 minutes
Number of questions–28

A CALCULATOR MAY NOT BE USED ON THIS PART OF THE EXAMINATION

Directions: Solve each of the following problems, using the available space for scratchwork. After examining the form of the choices, decide which is the best of the choices given and fill in the box. Do not spend too much time on any one problem.

In this test:

(1) Unless otherwise specified, the domain of a function f is assumed to be the set of all real numbers x for which $f(x)$ is a real number.

(2) The inverse of a trigonometric function f may be indicated using the inverse function notation f^{-1} or with the prefix "arc" (e.g., $\sin^{-1} x = \arcsin x$).

---

1.   What is the x-coordinate of the point of inflection on the graph of $y = xe^x$?

   (A) –2   (B) –1   (C) 0   (D) 1   (E) 2

Ans

---

2.   The graph of a piecewise-linear function f, for $-1 \le x \le 4$, is shown in the figure. If the function H is defined by

Graph of f

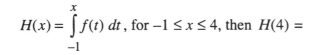$$H(x) = \int_{-1}^{x} f(t)\, dt, \text{ for } -1 \le x \le 4, \text{ then } H(4) =$$

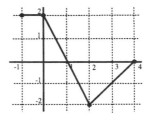

   (A) –2   (B) –1   (C) 0   (D) 1   (E) 2

Ans

---

3. $\int_{0}^{2} |x - 1|\, dx =$

(A) 0

(B) 1

(C) $\dfrac{1}{2}$

(D) 2

(E) 3

Ans

4. The function f is continuous at the point $(c, f(c))$. Which of the following statements could be false?

(A) $\lim\limits_{x \to c} f(x)$ exists

(B) $\lim\limits_{x \to c} f(x) = f(c)$

(C) $\lim\limits_{x \to c^{-}} f(x) = \lim\limits_{x \to c^{+}} f(x)$

(D) $f(c)$ is defined

(E) $f'(c)$ exists

Ans

5. $\int_{0}^{x} 2 \sec^{2}\left(2t + \dfrac{\pi}{4}\right) dt =$

(A) $2 \tan\left(2x + \dfrac{\pi}{4}\right)$

(B) $2 \tan\left(2x + \dfrac{\pi}{4}\right) - 2$

(C) $\tan\left(2x + \dfrac{\pi}{4}\right) - 1$

(D) $2 \sec\left(2x + \dfrac{\pi}{4}\right) \tan\left(2x + \dfrac{\pi}{4}\right)$

(E) $\sec\left(2x + \dfrac{\pi}{4}\right) \tan\left(2x + \dfrac{\pi}{4}\right)$

Ans

6. If $xy + x^2 = 6$, then the value of $\dfrac{dy}{dx}$ at $x = -1$ is

(A) -7 (B) -2 (C) 0 (D) 1 (E) 3

Ans

7. $\displaystyle\int_{2}^{3} \dfrac{x}{x^2 + 1}\, dx =$

(A) $\dfrac{1}{2}\ln\dfrac{3}{2}$ (B) $\dfrac{1}{2}\ln\dfrac{1}{2}$ (C) $\dfrac{1}{2}\ln 2$ (D) $2\ln 2$ (E) $\ln 2$

Ans

8. Suppose the function f is defined so that $f(0) = 1$ and its derivative, f', is given by $f'(x) = e^{\sin x}$. Which of the following statement are TRUE?

I $f''(0) = 1$

II   The line $y = x + 1$ is tangent to the graph of f at $x = 0$.

III   If $h(x) = f(x^3 - 1)$, then h is increasing for all real numbers x.

(A) I only   (B) II only   (C) III only   (D) I and II only   (E) I, II, III

Ans

9. Water flows into a tank at a rate shown in the figure. Of the following, which best approximates the total number of gallons in the tank after 6 minutes?

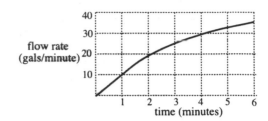

(A) 75   (B) 95   (C) 115   (D) 135   (E) 155

Ans

10. What is the instantaneous rate of change at $x = 0$ of the function f given by $f(x) = e^{2x} - 3\sin x$?

(A) –2   (B) –1   (C) 0   (D) 4   (E) 5

Ans

11. Suppose f is a function with continuous first and second derivatives on the closed interval $[a, c]$. If the graph of its derivative f' is given in the figure, which of the following is true?

(A) f is increasing on the interval (a, b)

(B) f has a relative maximum at $x = b$.

(C) f has an inflection point at $x = b$.

(D) The graph of f is concave down on the interval (a, b).

(E) $\int_a^c f'(x)\,dx = f(c) - f(a)$

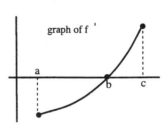

Ans

12. Suppose $F(x) = \displaystyle\int_0^{x^2} \dfrac{1}{2+t^3}\,dt$ for all real x, then $F'(-1) =$

(A) 2   (B) 1   C) $\dfrac{1}{3}$ (D) –2   (E) $-\dfrac{2}{3}$

Ans

☐

---

13. The graph of the function f is shown in the figure. For what values of x, $-2 < x < 4$, is f not differentiable?

(A) 0 only

(B) 0 and 2 only

(C) 2 and 3 only

(D) 0 and 3 only

(E) 0, 1 and 3 only

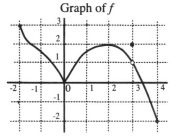

Graph of f

Ans

☐

---

14. A particle moves along the x-axis so that its position at any time $t \geq 0$ is given by $x(t) = \dfrac{t}{t^2 + 4}$. The particle is at rest when $t =$

(A) 0   (B) $\dfrac{1}{4}$ (C) 1   (D) 2   (E) 4

Ans

☐

15. Find the maximum value of $f(x) = 2x^3 + 3x^2 - 12x + 4$ on the closed interval [0,2].

(A) –3

(B) 2

(C) 4

(D) 8

(E) 24

Ans

16. If $f(x) = \ln(\cos 2x)$, then $f'(x) =$

(A) $-2\tan 2x$　　　(B) $\cot 2x$　　　　(C) $\tan 2x$　　　　(D) $-2\cot 2x$　　　　(E) $2\tan 2x$

Ans

17. The slope field for a differential equation $\frac{dy}{dx} = f(x,y)$ is given in the figure. The slope field corresponds to which of the following differential equations?

(A) $\frac{dy}{dx} = x + y$

(B) $\frac{dy}{dx} = -y$

(C) $\frac{dy}{dx} = y - \frac{1}{2}y^2$

(D) $\frac{dy}{dx} = x^2 + y^2$

(E) $\frac{dy}{dx} = y^2$

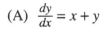

Ans

18. The *y*-intercept of the tangent line to the curve $y = \sqrt{x+3}$ at the point $(1, 2)$ is

   (A) $\dfrac{1}{4}$ (B) $\dfrac{1}{2}$ (C) $\dfrac{3}{4}$ (D) $\dfrac{5}{4}$ (E) $\dfrac{7}{4}$

Ans

19. The function defined by $f(x) = (x-1)(x+2)^2$ has inflection points at $x =$

   (A) -2 only

   (B) -1 only

   (C) 0 only

   (D) -2 and 0 only

   (E) -2 and 1 only

Ans

20. If $\displaystyle\int_0^b (4bx - 2x^2)\,dx = 36$, then $b =$

   (A) -6

   (B) -3

   (C) 3

   (D) 6

   (E) 15

Ans

21. If $\dfrac{dy}{dx} = -10y$ and if $y = 50$ when $x = 0$, then $y =$

(A) $50e^x$

(B) $50e^{10x}$

(C) $50e^{-10x}$

(D) $50 - 10x$

(E) $50 - 5x^2$

Ans

22. If $f(x) = x^3 - 5x^2 + 3x$, then the graph of f is decreasing and concave down on the interval

(A) $\left(0, \dfrac{1}{3}\right)$ (B) $\left(\dfrac{1}{3}, \dfrac{2}{3}\right)$ (C) $\left(\dfrac{1}{3}, \dfrac{5}{3}\right)$ (D) $\left(\dfrac{5}{3}, 3\right)$ (E) $(3, \infty)$

Ans

23. The figure shows the graph of f', the derivative of a function f. The domain of f is the closed interval $[-3, 4]$. Which of the following is true?

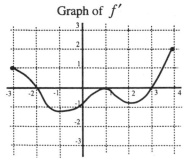

Graph of f'

    I. f is increasing on the interval $(2, 4)$.

    II. f has a relative minimum at $x = -2$.

    III. The f-graph has an inflection point at $x = 1$.

(A) I only

(B) II only

(C) III only

(D) I and II only

(E) I, II, III

Ans

24. How many critical values does the function $f(x) = \arctan(2x - x^2)$ have?

    (A) 0

    (B) 1

    (C) 2

    (D) 3

    (E) 4

    Ans

25. Which of the following is continuous at $x = 1$?

    I. $f(x) = |x - 1|$

    II. $f(x) = e^{x-1}$

    III. $f(x) = \ln(e^{x-1} - 1)$

    (A) I only

    (B) II only

    (C) I and II only

    (D) II and III only

    (E) I, II, III

    Ans

26. The number of motels per mile along a 5 mile stretch of highway approaching a city is modeled by the function $m(x) = 11 - e^{0.2x}$, where x is the distance from the city in miles. The approximate number of motels along that stretch of highway is

    (A) 16

    (B) 26

    (C) 36

    (D) 46

    (E) 56

    Ans

27. The diagonal z of the rectangle at the right is increasing at the rate of 2 cm/sec and $\dfrac{dy}{dt} = 3\dfrac{dx}{dt}$. At what rate is the length x increasing when $x = 3$ cm and $y = 4$ cm ?

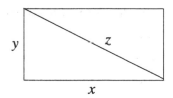

(A) 1 cm/sec

(B) $\dfrac{3}{4}$ cm/sec

(C) $\dfrac{2}{3}$ cm/sec

(D) $\dfrac{1}{3}$ cm/sec

(E) $\dfrac{1}{15}$ cm/sec

Ans

28. If $f(x) = \sin(2x) + \ln(x+1)$, then $f'(0) =$

(A) –1   (B) 0   (C) 1   (D) 2   (E) 3

Ans

EXAM VI
CALCULUS AB
SECTION I PART B
Time–50 minutes
Number of questions–17

A GRAPHING CALCULATOR IS REQUIRED FOR SOME QUESTIONS ON THIS PART OF THE EXAMINATION

Directions: Solve each of the following problems, using the available space for scratchwork. After examining the form of the choices, decide which is the best of the choices given and fill in the box. Do not spend too much time on any one problem.

In this test:

(1)   The <u>exact</u> numerical value of the correct answer does not always appear among the choices given. When this happens, select from among the choices the number that best approximates the exact numerical value.

(2)   Unless otherwise specified, the domain of a function f is assumed to be the set of all real numbers x for which $f(x)$ is a real number.

(3)   The inverse of a trigonometric function f may be indicated using the inverse function notation f^{-1} or with the prefix "arc" (e.g., $\sin^{-1} x = \arcsin x$).

---

1.   The graph of a function f is shown to the right.
       Which of the following statements about f is false?

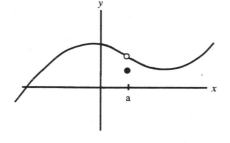

(A) f has a relative minimum at $x = a$.

(B) $\lim\limits_{x \to a^-} f(x) = \lim\limits_{x \to a^+} f(x)$

(C) $\lim\limits_{x \to a} f(x) \neq f(a)$

(D) $f(a) > 0$

(E) $f'(a) < 0$

Ans

☐

---

2.   The function f defined by $f(x) = e^{3x} + 6x^2 + 1$ has a horizontal tangent at $x =$

(A) -0.144 (B) -0.150 (C) -0.156 (D) -0.162 (E) -0.168

Ans

☐

---

3. Boyle's Law states that if the temperature of a gas remains constant, then the pressure P and the volume V of the gas satisfy the equation $PV = c$ where c is a constant. If the volume is decreasing at the rate of 10 in^3 per second, how fast is the pressure increasing when the pressure is 100 lb/in^2 and the volume is 20 in^3?

(A) $5\,\dfrac{\text{lb/in}^2}{\text{sec}}$ (B) $10\,\dfrac{\text{lb/in}^2}{\text{sec}}$ (C) $50\,\dfrac{\text{lb/in}^2}{\text{sec}}$ (D) $200\,\dfrac{\text{lb/in}^2}{\text{sec}}$ (E) $500\,\dfrac{\text{lb/in}^2}{\text{sec}}$

Ans

4. The graph of the second derivative of a function g is shown in the figure. Use the graph to determine which of the following are true.

I. The g-graph has points of inflection at $x = 1$ and $x = 3$.

II. The g-graph is concave down on the interval $(3, 4)$.

III. If $g'(0) = 0$, g is increasing at $x = 2$.

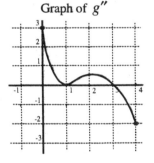

Graph of g''

(A) I only

(B) II only

(C) II and III only

(D) I and II only

(E) I, II, III

Ans

5. A particle moves along a straight line with its position at any time $t \geq 0$ given by

$$s(t) = \int_0^t (x^3 - 2x^2 + x)\, dx,$$ where s is measured in meters and t in seconds. The maximum velocity attained by the particle on the interval $0 \leq t \leq 3$ is

(A) 0.333 m/sec

(B) 0.148 m/sec

(C) 1 m/sec

(D) 3 m/sec

(E) 12 m/sec

Ans

6. If $\dfrac{dy}{dx} = \sqrt{2x+1}$, then the average rate of change y with respect to x on the closed interval $[0, 4]$ is

(A) 13   (B) $\dfrac{9}{2}$ (C) $\dfrac{13}{2}$ (D) $\dfrac{13}{6}$ (E) $\dfrac{1}{9}$

Ans
☐

7. If f' is a continuous function on the closed interval $[0, 2]$ and $f(0) = f(2)$,

then $\displaystyle\int_{0}^{2} f'(x)\, dx =$

(A) 0

(B) 1

(C) 2

(D) 3

(E) 4

Ans
☐

8.   If $k \neq 0$, then $\displaystyle \lim_{x \to k} \frac{x^2 - k^2}{x^2 - kx} =$

   (A) 0

   (B) 2

   (C) $2k$

   (D) $4k$

   (E) nonexistent

Ans

9.   Suppose that, during the first year after its hatching, the weight of a duck increases at a rate proportional to its weight. The duckling weighed 2 pounds when it was hatched and 3.5 pounds at age 4 months. How many pounds will the bird weigh at age 6 months?

   (A) 4.2 lbs

   (B) 4.6 lbs

   (C) 4.8 lbs

   (D) 5.6 lbs

   (E) 6.5 lbs

Ans

10. Let R be the region in the first quadrant enclosed by the x-axis and the graph of $y = \ln x$ from $x = 1$ to $x = 4$. If the Trapezoid Rule with 3 subdivisions is used to approximate the area of R, the approximation is

(A) 4.970   (B) 2.510   (C) 2.497   (D) 2.485   (E) 2.473

Ans

11. A solid has as its base the region enclosed by the graph of $y = \cos x$ and the x-axis between $x = -\frac{\pi}{2}$ and $x = \frac{\pi}{2}$. If every cross section perpendicular to the x-axis is a square, the volume of the solid is

(A) $\frac{\pi}{4}$

(B) $\frac{\pi^2}{4}$

(C) $\frac{\pi}{2}$

(D) $\frac{\pi^2}{2}$

(E) 2

Ans

12. If the function f is differentiable at the point $(a, f(a))$, then which of the following are true?

I. $f'(a) = \lim\limits_{h \to 0} \dfrac{f(a+h) - f(a)}{h}$

II. $f'(a) = \lim\limits_{h \to 0} \dfrac{f(a) - f(a-h)}{h}$

III. $f'(a) = \lim\limits_{h \to 0} \dfrac{f(a+h) - f(a-h)}{2h}$

(A) I only

(B) I and II only

(C) I and III only

(D) II and III only

(E) I, II, III

Ans

13. The level of air pollution at a distance x miles from a tire factory is given by

$$L(x) = e^{-0.1x} + \frac{1}{x^2}.$$

The average level of pollution between 15 and 25 miles from the factory is

(A) 0.144

(B) 0.250

(C) 0.156

(D) 0.162

(E) 0.168

Ans

14. Suppose the continuous function f is defined on the closed interval $[0, 3]$ such that its derivative f' is defined by $f'(x) = e^x \sin(x^2) - 1$. Which of the following are true about the graph of f?

   I. f has exactly one relative maximum point.
   II. f has two relative minimum points.
   III. f has two inflection points.

   (A) I only

   (B) II only

   (C) III only

   (D) I and II only

   (E) I, II, III

Ans

15. If the average value of $y = x^2$ over the interval $[1, b]$ is $\dfrac{13}{3}$, then the value of b could be

   (A) $\dfrac{7}{3}$ (B) 3   (C) $\dfrac{11}{3}$ (D) 4   (E) $\dfrac{13}{3}$

Ans

16. If the function f is defined on the closed interval $[0, 3]$ by $f(x) = \dfrac{2x}{x^2 + 1}$, which of the
following is true?

    I. $\displaystyle\int_0^3 f(x)\, dx = \ln 10$

    II. f has a relative maximum at $x = 1$.

    III. $f'(2) = \dfrac{1}{2}$

  (A) I only

  (B) II only

  (C) I and II only

  (D) II and III only

  (E) I, II, III

Ans

17. The area of the region bounded by the graphs of $y = \arctan x$ and $y = 4 - x^2$ is
approximately

  (A) 10.80

  (B) 10.97

  (C) 11.14

  (D) 11.31

  (E) 11.48

Ans

EXAM VI
CALCULUS AB
SECTION II, PART A
Time–45 minutes
Number of problems–3

A graphing calculator is required for some problems or parts of problems.

- Before you begin Part A of Section II, you may wish to look over the problems before starting to work on them. It is not expected that everyone will be able to complete all parts of all problems and you will be able to come back to Part A (without a calculator), if you have time after Part B. All problems are given equal weight, but the parts of a particular solution are not necessarily given equal weight.

- You should write all work for each problem in the space provided. Be sure to write clearly and legibly. If you make an error, you may save time by crossing it out rather than trying to erase it. Erased or crossed out work will not be graded.

- SHOW ALL YOUR WORK. Clearly label any functions, graphs, tables, or other objects you use. You will be graded on the correctness and completeness of your methods as well as your final answers. Answers without supporting work may not receive credit.

- Justifications require that you give mathematical (noncalculator) reasons.

- You are permitted to use your calculator in Part A to solve an equation, find the derivative of a function at a point, or calculate the value of a definite integral. However, you must clearly indicate in your exam booklet the setup of your problem, namely the equation, function, or integral you are using. If you use other built-in features or programs, you must show the mathematical steps necessary to produce your results.

- Your work must be expressed in mathematical notation rather than calculator syntax. For example, $\int_{1}^{5} x^2\, dx$ may not be written as $\text{fnInt}(X^2, X, 1, 5)$.

- Unless otherwise specified, answers (numeric or algebraic) need not be simplified.

- If you use decimal approximations in your calculations, you will be graded on accuracy. Unless otherwise specified, your final answers should be accurate to three places after the decimal point.

- Unless otherwise specified, the domain of a function f is assumed to be the set of all real numbers x for which $f(x)$ is a real number.

THE EXAM BEGINS ON THE NEXT PAGE

PLEASE TURN OVER

1.　Let R be the region in the first quadrant under the graph of $y = \dfrac{8}{\sqrt[3]{x}}$ for $1 \le x \le 8$.

(a) Find the area of R.

(b) The line $x = k$ divides the region R into two regions. If the part of region R to the left of the line is $\dfrac{5}{12}$ of the area of the whole region R, what is the value of k ?

(c) Find the volume of the solid whose base is the region R and whose cross sections cut by planes perpendicular to the x-axis are semicircles.

2. A particle starts at the point $(1,0)$ at $t = 0$ and moves along the x-axis so that at time $t \geq 0$ its velocity $v(t)$ is given by $v(t) = 1 + \dfrac{t}{1 + t^2}$.

   (a) Determine the maximum velocity of the particle. Show your work.

   (b) Find an expression for the position $s(t)$ of the particle at time t.

   (c) What is the limiting value of the velocity as t increases without bound?

   (d) Determine for which values of t, if any, the particle reaches the point $(101, 0)$.

3. The rate at which an air-conditioning unit for a theater complex pumps out cool air, in metric tons per hour, is given by a differentiable function R of time t. The table shows the rate as measured every hour over an 8-hour time period.

t (hours)	$R(t)$ (metric tons per hour)
0	4.6
1	5.4
2	6.1
3	6.5
4	6.8
5	6.3
6	6.1
7	5.5
8	4.8

(a) Use a midpoint Riemann sum with 4 subintervals of equal length to approximate $\int_0^8 R(t)\,dt$. Explain, using correct units, the meaning of your answer in terms of air flow.

(b) Is there some time t, $0 < t < 8$, such that $R'(t) = 0$? Explain.

(c) The rate of air flow $R(t)$ can be approximated using $Q(t) = \frac{1}{8}\left(36 + 8t - t^2\right)$. Use $Q(t)$ to approximate the average rate of air flow during the 8-hour time period.

A CALCULATOR MAY **NOT** BE USED ON THIS PART OF THE EXAMINATION.
DURING THE TIMED PORTION FOR PART B, YOU MAY GO BACK AND CONTINUE TO WORK
ON THE PROBLEMS IN PART A WITHOUT THE USE OF A CALCULATOR.

4.   The graph below shows the velocity v in feet per second of a small rocket that was fired from
       the top of a tower at time $t = 0$ (t in seconds). The rocket accelerated with constant upward
       acceleration until its fuel was expended, then fell back to the ground at the foot of the tower.
       The entire flight lasted 14 seconds.

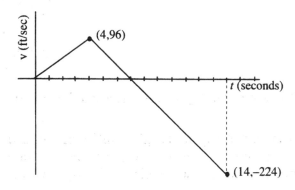

(a) What was the acceleration of the rocket while its fueled lasted?

(b) How long was the rocket rising?

(c) What was the maximum height above the ground that the rocket reached?

(d) How high was the tower from which the rocket was fired?

5.   Consider the differential equation $\frac{dy}{dx} = 2x - y$.

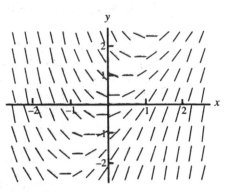

    (a) The slope field for the given differential equation is
         provided. Sketch the solution curve that passes
         through the point $(1, 0)$ and sketch the solution
         curve that passes through the point $(0, 1)$.

    (b) Find the value of b for which $y = 2x + b$ is a solution to the differential equation. Justify
         your answer.

    (c) Let g be the function that satisfies the given differential equation with the initial condition
         $g(0) = 0$. Does the graph of g have a local extremum at the point $(0, 0)$? If so, is the
         point a local maximum or a local minimum? Justify your answer.

    (d) Show that if C is a constant, then $y = Ce^{-x} + 2x - 2$ is a solution of the differential
         equation

6.   Let $G(x) = \int\limits_{-3}^{x} f(t)\, dt$ and $H(x) = \int\limits_{2}^{x} f(t)\, dt$ where f is the function graphed below.

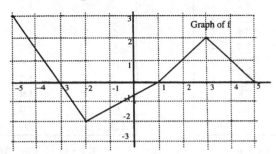

(a) How are the values of $G(x)$ and $H(x)$ related? Give a geometric explanation of this relationship.

(b) On which intervals of $[-5, 5]$, if any, is H increasing?

(c) At what x-coordinates, $-5 < x < 5$, does G have a relative maximum? Justify your answer.

(d) On which subintervals of $[-5, 5]$, if any, is G concave up?

Answers

EXAM I SECTION I PART A

1. A	11. A	21. C
2. B	12. D	22. B
3. E	13. C	23. D
4. C	14. B	24. D
5. E	15. B	25. B
6. D	16. E	26. B
7. A	17. C	27. C
8. D	18. C	28. E
9. B	19. B	
10. B	20. A	

EXAM II SECTION I PART A

1. C	11. C	21. D
2. A	12. B	22. A
3. B	13. A	23. C
4. C	14. D	24. B
5. D	15. A	25. B
6. C	16. C	26. E
7. E	17. C	27. A
8. A	18. E	28. E
9. B	19. B	
10. D	20. D	

EXAM I SECTION I PART B

1. D	7. B	13. C
2. C	8. B	14. C
3. B	9. C	15. B
4. E	10. C	16. D
5. D	11. E	17. E
6. D	12. E	

EXAM II SECTION I PART B

1. E	7. E	13. D
2. D	8. A	14. D
3. D	9. A	15. B
4. C	10. A	16. C
5. C	11. D	17. D
6. A	12. A	

EXAM I SECTION II PART A

1. a) -3 b) $\frac{3}{2}$ c) i) 0 ii) $-\frac{9}{2}$ d) 4

2. a) $(1, 5)$ b) $(-1, 5)$

   c) $A = \int\limits_{-2}^{1} (-x+6-f(x))\,dx = 6.75$

   $B = \int\limits_{-1}^{1} (5-f(x))\,dx = 4$ $A - B = 2.75$

   $\frac{4}{2.75} = \frac{16}{11}$

3. a) $c < \frac{1}{2}$ b) $\frac{16}{9}$ c) $\frac{4}{9}\pi$

EXAM II SECTION II PART A

1. a) 1.764

   b) 30.460

   c) 3.671

2. a) $2\pi\sqrt{2}$ in/sec b) $8(\pi - 2)$in^2/sec

3. a) 0

   b) $\frac{1}{6}$

   c) $y = \frac{1}{6}(x - 1)$

   d) $\left(\frac{1}{4}, \infty\right)$

EXAM I SECTION II PART B

4. c) $y = x - 1$

   d) $y = x - 1 + Ce^{-x} \Rightarrow x - y = 1 - Ce^{-x}$

   $\frac{dy}{dx} = \frac{d(x - 1 + Ce^{-x})}{dx} = 1 - 0 + Ce^{-x}$

   $= 1 + Ce^{-x}$

5. a) $y = 2x - 5$ b) $x = 1$ c) 0.75
   d) $x = -3, -1, 3$ e) $x = 4$

6. a) $A = \sin(\text{Arccos } k) - k\,\text{Arccos } k$

   b) $\frac{\sqrt{3}}{2} - \frac{\pi}{6}$ c) $\frac{dA}{dt} = -\frac{1}{3}$ units^2/sec

EXAM II SECTION II PART B

4. a) $(0, 1/10)$ has an inflection point
   b) $y = 4x + 1$

5. a) $(-3, 1)$ and $(3, 4)$

   b) Absolute max at $x = 1$

   c) $(-2, 2)$

6. a) $x < -\sqrt{\frac{b}{a}}$ or $x > \sqrt{\frac{b}{a}}$

   b) Rel max at $\left(-\sqrt{\frac{b}{a}}, -2\sqrt{ab}\right)$

   Rel min at $\left(\sqrt{\frac{b}{a}}, 2\sqrt{ab}\right)$

   c) $x > 0$ d) no inflection point.

Answers

EXAM III SECTION I PART A

1.	B	11.	A	21.	D
2.	C	12.	A	22.	C
3.	E	13.	C	23.	E
4.	E	14.	D	24.	A
5.	C	15.	B	25.	D
6.	C	16.	E	26.	B
7.	D	17.	C	27.	C
8.	B	18.	C	28.	E
9.	D	19.	A		
10.	E	20.	D		

EXAM IV SECTION I PART A

1.	A	11.	E	21.	B
2.	E	12.	B	22.	C
3.	D	13.	D	23.	C
4.	B	14.	D	24.	A
5.	C	15.	C	25.	A
6.	C	16.	E	26.	C
7.	E	17.	D	27.	D
8.	A	18.	D	28.	D
9.	D	19.	A		
10.	B	20.	D		

EXAM III SECTION I PART B

1.	A	7.	C	13.	E
2.	B	8.	D	14.	C
3.	B	9.	C	15.	A
4.	D	10.	C	16.	B
5.	E	11.	C	17.	E
6.	C	12.	B		

EXAM IV SECTION I PART B

1.	D	7.	B	13.	B
2.	D	8.	B	14.	C
3.	E	9.	D	15.	B
4.	D	10.	D	16.	D
5.	B	11.	C	17.	B
6.	C	12.	B		

EXAM III SECTION II PART A

1. a) 1.146

   b) $\displaystyle\int_0^a \pi(9\cos^2 x - e^{2x^2})\,dx$

   c) $\displaystyle\int_0^a (3\cos x - e^{x^2})^2\,dx$

2. a) $(0,0)$, $(0.964, 0)$, $(1.684, 0)$

   b) $(0, 0.398)$, $(1.351, 3)$

   c) min $= -0.098$, max $= 1.366$

3. a) 8 units²   b) $\dfrac{2\sqrt{3}}{3} \approx 1.155$

EXAM IV SECTION II PART A

1. a) $\dfrac{25}{12} \approx 2.083$ b) $\ln 5 \approx 1.609$

   c) $\sqrt{5} - 1 \approx 1.236$ c) 0.80

2. a) decreasing at 1 ft/sec

   b) distance is incr. at 1.471 ft/sec

   c) closest to balloon at $t = 44$ sec

3. a) $y = 110e^{-0.1t} + 70$

   b) 110.467°   c) $t > 7.885$ min

EXAM III SECTION II PART B

4. a) $p = -2$, $q = 5$ b) $p = -6$

   c) $p^2 < 3q$

5. a) $s(6) = 3$ km   b) only at $t = 4$ min

6. a) $\dfrac{y}{y^2 - x}$

   b) At $(1,2)$, $\dfrac{dy}{dx} = \dfrac{2}{3}$

   Tangent line: $y - 2 = \dfrac{2}{3}(x - 1)$; $y(1.3) \approx 2.2$

   c) $\dfrac{d^2y}{dx^2} = -\dfrac{4}{27}$

   d) overestimate

EXAM IV SECTION II PART B

4. a) i) $x = \dfrac{7}{6}\pi$ ii) $x = 0$

   b) $\left(\dfrac{\pi}{2}, \dfrac{3}{2}\pi\right)$ c) π

5. a) even   b) $\dfrac{4}{25}$ c) $1 - \dfrac{\pi}{4}$

   d) 1   e) $-1 < x < 1$

6. a) 0   b) 2

   c) $(-4, -3)$ and $(-1, 2)$

   d) $x = 1$

Answers

EXAM V SECTION I PART A

1.	B	11.	B	21.	B
2.	C	12.	D	22.	C
3.	C	13.	D	23.	D
4.	C	14.	A	24.	D
5.	D	15.	C	25.	B
6.	A	16.	B	26.	D
7.	E	17.	D	27.	A
8.	D	18.	E	28.	A
9.	A	19.	B		
10.	A	20.	B		

EXAM VI SECTION I PART A

1.	A	11.	E	21.	C
2.	C	12.	E	22.	C
3.	B	13.	D	23.	C
4.	E	14.	D	24.	B
5.	C	15.	D	25.	C
6.	A	16.	A	26.	D
7.	C	17.	A	27.	C
8.	E	18.	E	28.	E
9.	D	19.	B		
10.	B	20.	C		

EXAM V SECTION I PART B

1.	E	7.	D	13.	C
2.	C	8.	B	14.	E
3.	B	9.	E	15.	D
4.	E	10.	B	16.	D
5.	D	11.	E	17.	B
6.	C	12.	D		

EXAM VI SECTION I PART B

1.	E	7.	A	13.	A
2.	C	8.	B	14.	D
3.	C	9.	B	15.	B
4.	C	10.	D	16.	C
5.	E	11.	C	17.	B
6.	D	12.	E		

EXAM V SECTION II PART A

1. a) 8.3176

   b) $v(1) = e - \dfrac{1}{2} > 0$ so it moves to the right at $e - \dfrac{1}{2} \approx 2.218$ ft/sec.

   c) $t > 0.1756$ d) $x(.1756) = 0.7729$

2. a) $L = 39 \text{ m}^3, \ U = 49 \text{ m}^3$

   b) max error $= 5 \text{ m}^3$ c) 14.667 m^3

3. a) $2 - \dfrac{2}{e^k}$ units^2 b) $2\pi\left[1 - \dfrac{1}{e^{2k}}\right]$ units^3

   c) 2π units^3

EXAM VI SECTION II PART A

1. a) 36   b) 3.375   c) 24π

2. a) 1.5   b) $s(t) = 1 + t + \ln\sqrt{1+t^2}$

   c) 1   d) 95.441

3. a) 47.4 metric tons of air pumped out

   b) Yes. Since $R(2) = R(6)$, the MVT guarantees a time t between $t = 2$ and $t = 6$ where $R'(t) = 0$.

   c) 5.833 metric tons per hour

EXAM V SECTION II PART B

4. c) $y = C\sqrt{x^2 + 4}$ d) $y = 2\sqrt{x^2 + 4}$

5. a) $\lim\limits_{x \to -\infty} f(x) = 0$ $\lim\limits_{x \to \infty} f(x) = 0$

   b) $f'(x) = \dfrac{a(1-x^2)}{(x^2+1)^2}$; rel max $= \dfrac{a}{2}$

   rel min $= \dfrac{-a}{2}$ c) $\dfrac{\sqrt{2}}{\ln 3}$

6. a) $x < -1$ or $x > 0$ b) $\dfrac{1}{x(x+1)}$

   c) $y - \ln 0.5 = 0.5(x - 1)$ d) $\dfrac{e^x}{(1-e^x)^2}$

EXAM VI SECTION II PART B

4. a) 24 ft/sec^2 b) 7 seconds
   c) 784 ft   d) 448 ft

5. b) $b = -2$
   c) $g'(x) = 2x - y$ and $g'(0) = 0$
   $g''(x) = 2 - y' = 2 - 2x + y$
   at $(0, 0)$ $g''(0) = 2$ $\therefore$ rel. min.

6. a) The graph of H is the graph of G moved up 3.5 units.
   b) $(-5, -3)$ and $(1, 5)$
   c) $x = -3$
   d) $(-2, 3)$

Index

The *multiple-choice* questions and the *free-response* questions are listed in separate sections, each of which is divided into three broad categories:

I. Differentiation,
II. Integration,
III. Continuity, Limits & Graphs.

The problem type called *Graph Stem* refers to problems in which functions are defined graphically.

MULTIPLE-CHOICE
(Calculator-active problems are labeled **C**.)

I. DIFFERENTIATION

Applications p10#27; p109#7C

Average Rate Of Change p59#2C; p63#11C; p85#8C; p110#10C

Definition Of The Derivative & Theorems p11#1C; p16#15C; p26#3; p31#18; p35#1C; p37#7C; p50#5; p51#7; p75#6; p98#4; p114#16C; p124#10; p134#8C p136#12C

Graph Stems p2#3; p9#26; p11#1C; p12#5C; p19#1C p37#6C, #7C; p39#13C; p41#17C; p62#8C; p63#12C; p80#21; p83#1C; p84#3C; p86#9C; p87#12C; p104#21; p105#24; p113#15C; p121#2; p124#11; p125#13; p128#23; p131#1C; p132#4C

Implicit Differentiation p12#4C; p30#15; p34#28; p56#22; p74#5; p75#8; p105#25; p123#6

Increasing & Decreasing p1#1; p4#11; p10#28; p12#5C; p14#11C; p33#25; p41#17C; p56#23; p59#1C; p62#8C; p80#21,#23; p81#25; p83#1C; p84#3C; p86#9C; p102#17; p108#3C; p110#9C; p113#15; p128#22

Maximum & Minimum p2#5; p4#11; p13#6C; p32#22; p34#26; p39#13C; p40#15C; p55#20; p57#25; p59#1C; p78#15; p79#19; p86#9C; p99#6; p100#11; p101#13; p102#17; p108#3C; p110#8C; p113#14C; p126#15; p132#5C; p137#14C;

Product, Quotient & Chain Rules p1#2; p2#3, #5; p6#16; p7#19, #20; p8#22; p11#2C; p13#8C; p15#12; p28#11; p29#13; p30#16; p31#19; p32#21; p36#4C, #5C; p38#11C; p52#11; p54#15; p58#28; p65#15C; p74#3; p78#17; p97#1; p103#18,#20; p109#6C; p126#16; p129#24; p130#28; p134#8C; p138#16C

Related Rates p10#27; p37#6; p55#18; p63#12C; p76#10; p87#12C; p102#15; p109#7C; p130#28; p132#3C

2nd Derivative, Concavity & Inflection Points p3#7; p12#4C; p13#8C; p26#4; p39#13C; p40#14C; p41#17C; p51#8; p59#1C; p62#8C; p77#16; p80#21,#23; p83#1C; p84#3C; p86#9C; p99#6; p101#13; p102#17; p104#24; p107#1C; p110#8C,#9C; p113#14C; p121#1; p124#11; p127#19; p128#23; p132#4C

Slope Fields..p5#14; p22#2; p33#23; p53#14; p79#20..p97#2;..126#17

Tangents & Normals p1#2; p3#7; p7#20; p9#25; p12#5C; p25#1; p32#21; p36#3C; p39#13; p40#15C; p49#1; p53#13; page 56#22; p63#11C; p64#14C; p74#5; p75#7; p78#17; p84#5C; p86#10C; p87#13C; p98#3; p105#24,#25; p127#18; p131#2C

Velocity & Acceleration p9#24, #26; p14#11C; p17#16C; p27#6; p34#28; p58#27; p64#13C; p82#26; p85#6C; p104#23; p125#14; p132#5C

II. INTEGRATION

Applications p14#9C; p17#16; p35#2C; p63#12C; p81#24; p107#2C; p108#3C; p124#9; p129#26

Area p6#17; p13#7C; p26#5; p30#17; p37#6C; p39#12C; p50#4; p60#4C; p62#9C; p81#25; p85#7C; p86#10C; p105#26; p138#17C

Average Value p5#13; p33#24; p52#9; p76#11; p100#10; p136#13C; p137#15C

Definite Integrals p2#4; p8#21; p12#3C; p25#2; p27#8; p28#10; p29#14; p35#2C; p37#8C; p49#2; p51#6; p55#19; p56#21; p59#2C; p77#13; p89#16; p101#14; p102#16; p122#3, #5; p123#7; p127#20

Differential Equations p8#23; p14#9C; p29#12; p35#2C; p38#10C; p58#28; p65#16C; p81#24; p89#17C; p106#27; p107#2C; p128#21; p133#6C; p134#9C

Index

Fundamental Theorem p5#12; p17#17C; p34#27; p37#8C; p53#12; p54#17; p59#2C; p62#10C; p74#4; p82#28; p85#8C; p106#28; p111#11C; p113#15C; p121#2; p125#12; p133#7C

Graph Stems p14#10; p37#6C, #7C; p57#26; p61#6C; p63#11C; p89#16C; p104#21; p113#15; p124#9;

Numerical Integration p3#6; p14#10C; p37#7C; p57#26; p60#3C; p63#11C; p80#22; p86#11C; p108#4C; p112#13C; p135#10C

Symbolic Integration p3#8; p7#18; p15#13C; p27#7; p33#23; p73#2; p82#27; p84#4C; p101#12; p103#19

Velocity & Acceleration p14#11C; p17#16C; p41#16C; p52#10; p61#7C; p76#9; p99#7; p114#17C

Volume p4#10; p9#24; p14#10C; p38#9C; p57#24; p61#7C; p79#18; p88#15C; p108#5C; p135#11C

III. CONTINUITY, LIMITS, GRAPHS

Continuity p4#11; p16#15C; p31#20; p36#4C; p50#3; p62#8C; p77#14; p83#2C; p84#5C; p98#4; p100#9; p114#16C; p122#4; p129#25; p131#1C

Limits p4#9; p6#15; p11#1C; p26#3; p28#9; p36#4C; p50#5; p54#16; p60#5C; p73#1; p75#6; p77#12, #14;; p99#8; p100#9; p123#8

FREE-RESPONSE

(Problems #1, #2, and #3 are calculator-active. No calculators are used on #4, #5, and #6.)

I. DIFFERENTIATION

Average Rate Of Change p19#1; p116#1

Graph Stems p23#5; p47#5; p71#5; p94#4; p120#5; p145#6

Implicit Differentiation p72#6

Increasing & Decreasing p47#5; p48#6; p68#2; p70#4; p93#3; p143#4

Maximum & Minimum p47#5; p48#6; p68#2; p69#3; p94#4; p96#6; p120#5; p145#6

Product, Quotient & Chain Rules p95#5; p120#5; p121#6

Related Rates p24#6; p44#2; p92#2

2nd Derivative, Concavity & Inflection Points p20#2; p23#5; p47#5; p48#6; p70#4; p72#6; p96#6; p145#6

Tangents & Normals p19#1; p20#2; p69#3; p72#6: p121#6

Velocity & Acceleration p19#1; p71#5; p116#1; p143#4

II. INTEGRATION

Accumulation Functions p45#3C

Applications p91#1

Area p20#2; p21#3; p24#6; p43#1; p45#3; p67#1; p69#3; p91#1; p118#3; p140#1C

Average Value p19#1; p142#3C

Definite Integrals p95#5; p118#3

Differential Equations p91#3; p119#4

Graph Stems p96#6; p117#2

Numerical Integration p45#3; p94#4; p117#2; p

Slope Fields p22#4; p46#4; p118#4C; p144#5

Velocity & Acceleration p71#5; p141#2; p143#4

Volume p21#3; p43#1; p67#1; p91#1; p118#3; p140#1C

III. CONTINUITY, LIMITS, GRAPHS

Graphs p71#5; p120#5

Limits p94#4; p141#2C